Minor Spices and Condiments

Kodoth Prabhakaran Nair

Minor Spices and Condiments

Global Economic Potential

Springer

Kodoth Prabhakaran Nair
Villament G3
C/o Dr. Mavila Pankajakshy
Malaparamba, Kozhikode, Kerala, India

ISBN 978-3-030-82245-3 ISBN 978-3-030-82246-0 (eBook)
https://doi.org/10.1007/978-3-030-82246-0

This Springer imprint is published by the registered company Springer Nature Switzerland AG
The registered company address is: Gewerbestrasse 11, 6330 Cham, Switzerland

The late Dr. A. P. J. Abdul Kalam, India's great president, launched the book ***Issues in National and International Agriculture***, authored by Professor Kodoth Prabhakaran Nair, at the Raj Bhavan (Governor's Office), Chennai, Tamil Nadu, India

My late father Kuniyeri Pookkalam Kannan Nair, a well-respected police officer who served the Imperial Police and was decorated with the King George V medal for bravery and honesty, and my late mother Kodoth Padinhareveetil Narayani Amma, daughter of the aristocratic Kodoth family of North Malabar, Kerala State, India; I lost them both at a very young age, but whose endearing love and blessings made me what I am today

I dedicate this important book, compiled at the most difficult phase of my professional life, with so many constraints coming my way, in particular, due to the global covid pandemic, to the memory of my late eldest brother Dr. Kodoth Padinhareveetil Krishnan Nair, a Physician par excellence, a magnanimous soul, one who had helped innumerable persons, and his wife, Shrimathi Sathy V. Nair, who treated me like a mother, and, also my elder brother late Engineer Kodoth Padinhareveetil Rathnakaran Nair, known for his foresight and forth righteousness, who motivated me to study agricultural science, in India, at a time when the subject was least sought after, in fact, looked down upon. The former financed my undergraduate studies in agriculture, the latter, my post graduate studies. My late sisters, Kodoth Padinahareveetil Karthiayini Amma and Dr. Kodoth Padinhareveetil Meenakshi Amma were also supportive of my education, though not financially, during my very difficult younger days

Acknowledgements

It is my great pleasure to very sincerely thank Ms. Deepthi Vasudevan, Project Editor (Books), who was so very conscientious of her editorial responsibilities, and Mr. P. Balaji, Project Manager, and his team, who did a remarkable job in the production of this important book.

Dr. J. Rema, Director, Indian Institute of Spices Research (IISR), Calicut, Kerala State, India, was very gracious to provide most of the excellent color photographs. I am also grateful to Dr. Jayasree, Head of Post Harvest Technology at IISR, who also very kindly provided some of the color photos. I am much indebted to both.

Contents

Introduction 1
References 2

Onion and Garlic -Bulbous Spices 3
Onion 3
Varieties 4
New Varieties 5
Crop Management Practices 5
Soil Preparation 5
Crop Rotation 6
Propagation 6
Transplanting 6
Direct Bulb Planting for Green Onions as Salad 7
Direct Planting Method 7
Irrigation 8
Interculture, Weeding and Rogueing 8
Nutrition 8
Harvest 9
Yield 9
Seed Production 9
Isolation Distances 10
Crop Protection 10
Diseases 11
Marketing 11
Storage 12
Cold Storage 13
Irradiation 13
Effect of Nitrogen Fertilization on Storage Behavior 14
Curing 14
Neck Cutting 14
Sorting and Grading 14

Packaging 15
Effect of Packing Material on Quality 15
Composition 20
Reference 21

Garlic 23
Crop Management 25
Marketing 26
Curing, Packaging and Storing 27
Composition 31
Uses of Garlic 34
Garlic as an Insecticide 35
Garlic in Medicine 35
References 36

Leek/Stone Leek/Welsh Onion 39
Crop Management 40
Reference 41

Ajowan or Bishop's Weed 43
Crop Management 44
Cultural Practices 44
Rainfed Cultivation 45
Post-Harvest Technology 45
Processed Products 46
Essential Oil 46
Medical Use 47
References 48

Celery Seed 49
Economic Importance and International Trade 50
Area and Production 50
Crop Management 50
Land Preparation 51
Sowing Method 51
Fertilizers 51
Irrigation 51
Plant Protection 52
Tissue Culture 52
Threshing and Harvesting 52
Post Harvest Technology 52
Adulteration in Celery Seed 53
Grading 53
Packaging and Storage 54
Processed Products 54
Seed-Based Processed Products 54
Celery Seed Oleoresin 55

Processed Products from Celery Leaves and Stalks 56
Celery Stalk Products 56
How to Handle and Store Celery Seed Ad Its by Products? 57
Uses 57
Reference 58

Coriander 59
Description and Distribution 59
Origin and Brief History 59
Commercial Production 60
Botanical Description 60
Economic Importance and International Trade 60
Plant Protection Measures 61
Controlling Insect Pests 63
Additional Plant Protection Measures to Control Diseases and Pests 63
Harvesting and Threshing 64
Packaging and Storage 66
Natural Occurrence of Aflatoxins in Coriander Seeds 66
Processed Products 67
Coriander Dal – An Improved CFTRI Product 68
Seasoning from Coriander Roots 68
Uses of Coriander 69
References 69

Cumin Seed or Safaid Jeera 71
Description and Distribution 71
Brief History and Distribution 71
Area and Production 72
Crop Management 72
Cultivars 73
Sowing Time 74
Commercial Quality 74
Cumin Powder 75
Volatile Oil 75
New Supercritical Fluid Extraction (SCFE) Process for Volatile Oil 76
Uses 76
References 77

Fenugreek 79
Description and Distribution 79
Area, Production and Export 80
Crop Management 80
Climate and Soil 80
Manuring 80
Propagation 81
Seed Rate 81

Irrigation and Interculture 82
Tissue Culture 82
Plant Protection 82
Post-harvest Technology 83
Marketing 84
Grades and Specifications 84
Processed Products 85
Oleoresin 85
Uses 86
References 87

Mustard 89
Origin and Historical Background 89
Yellow or White Mustard 90
Brown and Oriental Mustard Seed 90
Black Mustad Seeds 91
The Uniqueness of Yellow Mustard 91
Description and Distribution 91
Area, Production and Export 92
Crop Management 92
Plant Protection 92
Post-harvest Technology 93
Quality Evaluation 93
Brown Mustard(*Brassica juncea*) 94
Quality Standards 94
Processed Products 95
Mayonnaise 95
Dried/Dehydrated Mustard Greens (Leaves) 96
Mustard for Industrial Buyer 96
Uses 97

Cassia and Cinnamon 99
Nomenclature 99
Species of Cinnamon Grown in India 100
Description and Distribution 100
Crop management/Propagation 101
Uses 101
Tejpat or "Tamal Patra" or Indian *Cassia lignea* 101
Description and Distribution 102
Crop Management 102
Harvesting and Collection of Tamala Leaves 102
Preparation for Marketing 103
Uses 103
Physico-chemical Specifications of Cassia Types 104
Packing 104
References 105

Cinnamon or True Cinnamon . . . 107
Description and Distribution . . . 107
Area, Production and Export . . . 108
World Trade in Cinnamon . . . 108
Global Market Outlook . . . 109
Crop Management . . . 109
Transplanting of Cinnamon into Polythene Bags . . . 110
Propagation of Cinnamon by Cuttings . . . 110
Transplanting Seedlings . . . 110
Manuring and Intercultivation . . . 110
Weeding . . . 111
Pruning/Cropping . . . 111
Variability Studies . . . 111
Plant Protection . . . 112
Harvesting . . . 113
Post Harvest technology: . . . 114
Preparation and Cutting of Bark Spice for Marketing: . . . 114
Cutting and Peeling . . . 114
Preparation of Quils . . . 114
Drying of Quils . . . 115
Bark Anatomy . . . 115
Utilization of Leaves and Twigs . . . 116
Economics of Cultivation . . . 116
Grading . . . 116
Type Seychelles . . . 117
Type Madagascar . . . 119
Packaging of Different Commercial Classes of Cinnamon Bark and Powder: . . . 120
Whole cinnamon: . . . 120
Cinnamon powder: . . . 120
Packing of Cinnamon and Cassia and their products: . . . 120
Storage of Cinnamon and Fumigation . . . 121
Transport of Cinnamon . . . 121
Processed Products . . . 121
Cinnamon Bark Oil . . . 121
Cinnamon Leaf Oil . . . 122
Seasonal Variation in Oil Recovery . . . 122
Physio-Chemical/Quality Specifications for Cinnamon Leaf Oil . . . 123
Physical Specifications . . . 123
Cinnamon and Cassia Oleoresins . . . 123
Other Products . . . 123
Packaging and Storage . . . 123
Uses . . . 124
Stem Bark Oil . . . 124
Fungitoxic Properties of Cinnamon Bark Oil . . . 124

Cinnamon Leaf Oil 125
Root Bark Oil 125
Seed Oil 125
Cinnamon Buds 125
Cinnamon Wood 126
References 126

Clove 127
Description, Distribution, and Economic Importance 127
Area, Production and Export/Import 128
What Are the Constraints of Clove Production in India? 128
Crop Management 128
Varieties 129
Propagation 130
Factors Affecting Germination 130
Raising Nursery 131
Transfer of Seedlings to Bamboo Baskets/Mud Pots/Polythene Bags 131
Transplanting in the Main Field 131
Propagation by Grafting 131
Mixed Cropping 131
Weeding 132
Support 132
Manuring 132
Interculture/Thinning 132
Soil and Moisture Conservation Work 132
Water Management 133
Harvesting and Curing 133
Plant Protection 134
  Insect Pests 134
  Animals 134
  Plant Diseases 134
  Nursery Pests and Diseases 135
Yield 135
Economics of Cultivation 135
Constraints in Clove Cultivation 135
Post Harvest Technology 136
  Preparation of Clove for the Market 136
  Grading 136
Grading of Whole Cloves 137
Physico-Chemical Quality Requirements/Specifications 137
Ground (Powdered) Cloves 137
Packing of Whole and Ground Cloves 137
Storage 138
Transport 138

Defective Cloves . . . 138
Composition . . . 139
Adulteration . . . 139
Quality Standards . . . 139
Packing of Cloves . . . 139
Packaging of Clove Oil . . . 139
Packaging of Clove Oleoresin . . . 140
Processed Products . . . 140
  Clove Bud Oil . . . 140
  Clove Stem Oil . . . 141
  Clove Leaf Oil . . . 141
  Oil of Mother Cloves . . . 141
  Clove Root Oil . . . 141
Uses . . . 142
References . . . 142

Mace and Nutmeg . . . 143
Description and Distribution . . . 143
Area, Production and Exports/Imports . . . 144
Crop Management . . . 144
Propagation by Nursery . . . 144
Standardization of Seed Propagation Technique for Nutmeg: Advancements . . . 145
Transplanting of Seedlings in the Field . . . 145
Manuring . . . 146
Mulching . . . 146
Interculture/Weeding . . . 146
Plant Protection . . . 146
Diseases . . . 147
Constraints in Nutmeg Cultivation . . . 147
Harvesting and Curing . . . 147
Yield . . . 148
Economics of Cultivation . . . 148
Post Harvest Technology . . . 148
  Packaging of Mace and Nutmeg . . . 148
Grading . . . 149
Nutmeg . . . 150
Processed Products . . . 151
  Mace Products . . . 151
Nutmeg Butter . . . 153
Uses . . . 154
  Nutmeg . . . 154
  Mace . . . 155
References . . . 155

Allspice or Pimenta or Pimento 157
Description and Distribution 157
Distribution 158
Crop Management 158
Propagation 158
Propagation by Seed/Nursery Seedlings 158
Viability of Allspice Seedlings 159
Propagation by Budding 159
Propagation by "Approach Grafting" 159
Top Working 160
Rooting in Allspice by Air Layering 160
Transplanting of Seedings 160
Weeding/After Care and Fertilizer Schedule 160
Successful Fruiting in Allspice 161
Harvesting and Yield 161
Major Types of Allspice (Pimento) 161
Plant Protection 162
  Diseases 162
  Insect Pests 162
Post Harvest Technology 163
  Grades/Types 163
International Organization for Standardization 163
  Storage of Pimento 163
  Transport of Pimento 164
  Packaging and Storage 164
Processed Products 164
  Pimenta Berry Oil 164
  Pimenta Leaf Oil 165
  Pimenta Oleroresin 165
  As Food Flavorant 165
  Use in Medicine 166
  Berry Oil 166
  Leaves 166
  Pimenta Bark and Wood 166
References 167

Curry Leaf 169
Description and Distribution 169
Area, Production and Exports 169
Crop Management 170
Curry Leaf Cultivation in Homelands of Southern Kerala, India 170
Raising Nursery 171
Manuring 171
Harvesting 171
Weeding 171

Prophylactic Measures 172
Cost of Cultivation 172
Plant Protection 172
Diseases 173
Post Harvest Technology 174
Processed Products 174
Volatile Oil 174
Dehydrated Curry Leaves 175
References 176

Tamarind 177
Description and Distribution 177
Area, Production and Exports 178
Crop Management 178
Soil and Climate 178
Varieties 178
Propagation 178
Patch Budding for Quicker Multiplication on Large Scale 179
Growth/Development of Seedlings 179
Irrigation 179
Manuring 179
After Cultivation 179
Disease Control 180
Insect Pests 180
Nematodes 180
Harvesting 180
Yield 181
Post Harvest Technology 181
Preparation for the Market 181
Composition 181
Tamarind Pulp 181
Tamarind Concentrate 182
Tamarind Powder 182
Grading 182
Processed Products 183
Uses 183
References 184

Asafoetida or Asafetida 185
Description and Distribution 185
Area, Production and Trade 186
Method of Gum-Resin Collection 187
Crop Management 187
Composition 187
Adulteration 188
Processed Products 188

Compounded Asafoetida Powder and Tablets: A Simple CFTRI Process . . . 189
Details of CFTRI Process . . . 189
Raw Materials Needed for 200 kg Batch . . . 189
Uses . . . 190
Some Important Concluding Observations . . . 190
References . . . 191

Vanilla . . . 193
Description and Distribution . . . 193
Commercial Forms of Vanilla . . . 194
Area, Production and Trade . . . 194
Varieties . . . 195
Future Prospects . . . 195
Major Constraints in Natural Vanilla Production . . . 195
A definite Future for Natural Vanilla . . . 196
Crop Management . . . 196
Manuring . . . 197
After-Care . . . 198
Flowering and Natural Pollination . . . 198
Hand Pollination . . . 198
Pod Development . . . 199
Tissue Culture for Vanilla Production . . . 199
Yield . . . 199
Plant Protection . . . 199
Diseases . . . 199
Pests . . . 200
Post-harvest Technology . . . 200
Natural Mechanism of Flavor (Vanillin) Generation During Curing . . . 201
Enzymic Generation of Vanillin . . . 201
Yield of Cured Beans . . . 203
Mechanized Method of Curing and Modification of the Vanilla Preparation Method: . . . 203
Vanilla Bean Packing and Storage . . . 203
Vanilla Pods . . . 204
Cut Vanilla . . . 204
Vanilla in Bulk . . . 204
Vanilla for Retail Trade . . . 204
Quality Attributes of Vanilla . . . 204
Some Precautions . . . 205
Adulteration . . . 206
Processed Products . . . 206
Vanilla Powder, "Vanilla Absolute" and Vanilla Tincture . . . 207
Uses . . . 207
As Food Flavorant and in Perfumery . . . 207
References . . . 226

Introduction

The historians Collins and Lapierre, in their monumental work "Freedom at Midnight" (1976) make a remarkable statement as follows.

"Sometimes history's most grandiose accomplishments have the most banal of origins. Great Britain was set on the road to the great colonial adventure for five miserable shillings. They represented the increase in price of a pound of Black pepper proclaimed by the Dutch privateers who then controlled the spice trade. Incensed at what they considered a wholly unwarranted gesture, 24 merchants of the city of London gathered on the afternoon of 24 September 1599 A.D. in a decrepit building on Leadenhall Street. Their purpose was to find a modest trading firm with an initial capital of 72,000 Sterling Pounds subscribed by 125 shareholders. Only the simplest of concerns, that is, profit, inspired their enterprise, which expanded and transformed, would ultimately become the most noteworthy creation of the age of imperialism – the British Raj". Verily, India's destiny was then charted.

There are many spices, which constitute an important group of agricultural commodities, which, since antiquity, have been considered indispensable in the culinary art for flavoring foods. Published work have considered many, such as, Back pepper, Cardamom, the former known as the "King" of spices, the latter as the "Queen" of spices, Turmeric, Ginger etc., and recent published work (Nair 2019, 2020a, b) have further updated the scientific information on these. But, on the whole, published work on minor spices is, but, scanty. Why spices are very important to Indian economy is because they earn a huge amount of hard currency through export. In fact, the economy of states like Kerala, in India, primarily depends on the earning of the above named spices, the major ones. This book will explore the enormous potential of minor spices and condiments on not just Indian economy, but, many others on the Asian and African continents.

Definition of spices and condiments: According to the latest available International Organization for Standardization (ISO) report, there are about 86 spices grown in different parts of the world. Of these, India grows 56. Both ISO and ISI (now BIS, Bureau of Indian Standards) experts concluded, after much

K. P. Nair, *Minor Spices and Condiments*,
https://doi.org/10.1007/978-3-030-82246-0_1

deliberation, that there is no clear-cut division between "spices" and condiments". As such, they have been clubbed together. The term "Spices and Condiments" applies to such natural plant or vegetable products or mixtures thereof, used in whole or ground form, mainly to impart flavor, aroma and piquancy to foods and also for seasoning of foods and beverages like soups and sauces. The following chapters sequentially discuss the various spice crops and the condiments.

References

Nair KPP (2019) Turmeri (*Curcuma longa* L.) and Ginger (Zingiber officinale Rosc.) – World's Invaluable Medicinal Spices. Springer Nature Switzerland

Nair KPP (2020a) The geography of Black Pepper (*Piper nigrum*) The "King" of Spices Vol 1. Springer Nature Switzerland

Nair KPP (2020b) The geography of cardamom (*Elettaria cardamomum* M.) The "Quee" of spices. Vol 2. Springer Nature Switzerland

Onion and Garlic -Bulbous Spices

Onion

Botanical name: *Allium cepa* Linn.
Family: Liliaceae

Brief History Onion is too well known to need any description. In the oldest recorded history onions were depicted as food in Egyptian tombs as early as 3200 BC. Onion is mentioned both in the Bible and Koran.

Types of Onions There are six types of onions, as follows:

1. Green onions which are used in salads
2. Mild onions which are used in cooking (preparation of Indian curry etc) as food or as a salad
3. Pungent varieties which are used as condiment for flavoring of a number of food items (primarily Indian)
4. Pearl onions or small onions which are used in pickles, including vinegar pickles
5. White or yellow onions for dehydration purpose and for the manufacture of onion powder; white onions of desired quality are preferred for this purpose.
6. Red and yellow onions are general purpose crops for culinary purpose

During 1994–1995 there were about 384, 400 hectares (ha) under onion, producing nearly 4057,900 metric tons (mt) with an Indian average yield 111 mt/ha. Maharashtra had the largest acreage (101, 600 ha), followed by Karnataka (66, 700 ha), Orissa (43,800 ha) and Uttar Pradesh (29,500 ha). Even now Maharashtra continues to lead all the states in acreage.

Export of Indian onion, probably, is the biggest hard currency earner for India. Indeed, it also plays a pivotal role in Indian export, since, staples like wheat, (most important for the whole of North Indians), without onion, either raw or in cooked form, is incomplete for a north Indian meal. Even onion seeds are exported from

K. P. Nair, *Minor Spices and Condiments*,
https://doi.org/10.1007/978-3-030-82246-0_2

India. The major market for Indian onion is the UAE (United Arab Emirates) along with Saudi Arabia, Singapore, Malaysia and Japan in the far east, while Sri Lanka and Bangladesh are also major importers, along with Russia.

Varieties

Possibly, there are 300 species of onion. *Allium cepa* or garden onion is possibly the most widely cultivated species. This species is further divided into the following three types, as follows:

1. Extensively cultivated bulbing type or common onion group, propagated by seed
2. Multiplier type or "Aggregation Group" propagated by branching at its base
3. Top onion type or "Proliferum Group" forming bulbils in the inflorescence, which are often used for propagation (Chaudhury 1983). Of these, the bulbing type onion is mostly cultivated commercially in India and constitutes the most important commercial bulb crop in India.

There are several varieties of onion available in the market. These are known by color, shape, size, pungency and shelf-life quality of the bulb. In India, the onion varieties can be divided into "pungent" and "mild" types. Three distinct colors are recognized in each class, such as, (1) red, (2) white and (3) yellow.

What are the quality parameters of a good variety of onion?

In general, an onion variety which is vastly accepted by people, should have the following qualities:

1. It should be of medium size
2. Tough in texture
3. Strong in flavor
4. Thin at the neck
5. Globular in shape
6. Should have good shelf-life under ordinary storage conditions
7. The variety should not have the tendency to bolt at the bulb-forming stage
8. There should be no double or triple bulb forming or branching

   Keeping the above-mentioned quality attributes, intensive research on varietal improvement has been going on in different Indian onion-growing states. The varieties grown are usually called after the names of the states, such as, Patna Red, Patna White, Nasik Red, Poona Red, Belari Red (Andhra Pradesh), Punjab Selection, Hissar II etc. Three varieties, namely, Early Grano, Pusa Red and Pusa Ratnar have been recommended by India's premier agricultural research institute, namely, Indian Agricultural Research Institute, situated in New Delhi. Early Grano (introduced from USA) has large, globular, yellow bulbs which are mildly pungent and are good for salads. Pusa Red is a medium sized, red colored onion which has a good shelf life. Pusa Ratnar is a large sized, high yielding, red colored onion. Regional field trials conducted in Bihar,

Maharashtra, Delhi and Punjab have shown its success and superiority over most other selections. Among the several white onion varieties tested, Raina, Pruthi, and associates recommended "Punjab Selection"widely grown in Punjab.

New Varieties

"N-257-9-1"(white) and "N-2-4-1" (red) developed at Rahuri were approved for release. The Indian Institute of Horticultural research, Bengaluru, Karnataka State, has released the following 3 varieties for the south Indian region. These are the following:

1. Arka Niketan, having 13–14% total soluble solids, yields about 42tons/ha
2. Arka Kalyan, yielding about 47tons/ha which is resistant to purple blotch
3. Arka Pragati, yielding about 45tons/ha which is suitable for cultivation both as a *Rabi* and *Kharif* crop

Crop Management Practices

Climatic Requirements

Cool weather is favorable in the early growing period. The young plants are quite tolerant to cold and withstand freezing to a certain degree. Onion needs dry atmosphere and moderately high temperature during maturity. Longer days and higher temperature promote flowering and bulb formation. Day length affects onion and increase in photoperiod hastens maturity.

Temperature (both ambient and soil) influences bulb formation. A range of 10–15.6 degree Celsius checks bulb formation, while, increase from 21.1 to 26.7 degree Celsius hastens bulb formation. The effect of temperature is more pronounced than the photoperiod on the development of seed stalks. At relatively low temperatures of 10–15.6 degree Celsius coupled with a short day of 9–12 h onion plant will move to seed formation steadily, while at higher temperature of 21.1–26.7 degree Celsius the plant will not seed either under a short day of 9–12 h or a long day of 15 h.

Soil Preparation

All types of soils are suited to onion. High yields, however, are only obtained in heavy soils. Clay soils are alright, but, must be well supplied with humus. Heavy clay soils lead to water logging. The most desirable soil for onion is that which retains sufficient moisture but at the same time is sufficiently favorable for cultivation, permitting proper development of bulbs. The most favorable soil pH for onion is 6.5–7.8.

Soils for onion production should be prepared well. Well-rotted farm yard manure should be applied at the rate of 15 t/ha following first ploughing so that it may become well mixed during subsequent operations. The seed beds should be well pulverized and should have a smooth surface since onion needs a fine tilth than most other crops.

In Punjab and Haryana seed is sown in nursery beds from October to November and the seedlings are transplanted from mid December to end January. In the hilly regions, the crop is raised during summer months (March to May/June) and the seed is sown in the nursery from end February to end May. Onion is planted up to an elevation of about 2000 m above mean sea level.

Crop Rotation

Onion crop should follow those vegetable crops which have been manured heavily. Cauliflower, tomato and potato are the ideal crops for this purpose as these require heavy applications of organic matter to the soil.

Propagation

Three methods of propagation (1) transplanting (2) planting bulbs directly in the field and (3) broadcasting or dribbling the seed in the prepared soil, are followed. Details follow:

Transplanting

The seeds are planted in the nursery beds, and the seedings thus raised are then carefully lifted and transplanted in the main field. Seed beds, 125cmn wide and 300 cm long raised to a Height of 10 cm are first prepared. Sufficient space is provided between two seed beds to facilitate weeding and hoeing. Seed beds are properly manured and levelled before seeding. The seeds are then broadcast uniformly or sown in rows and covered over lightly by well-rotted farmyard manure. Thick planting should be avoided. 10–12 kg seed is sufficient to plant a nursery for one hectare. Any insecticide (like BHC 5%) may be applied around the seedbed to prevent the entry of black ants. First watering should be done with a sprinkler and subsequent irrigations should be every third or fourth day until the plants have established well.

When the seedlings are 7–8 cm tall, very dilute solution of calcium nitrate should be evenly sprinkled after every 8–10 days. Immediately after this, water should be sprinkled to prevent leaf burning. The seedlings are ready to transplant in about 6–8 weeks time.

When the seedbed is ready, the main field is divided into small and convenient-sized plots. The seedlings, when large enough, are transplanted in rows, by placing them in furrows previously opened by a rake. The bulbs of the nursery plants are covered by putting the soil back in the furrow. They may be planted by dibbling when a hole is made by a wooden peg, and, the seedlings are planted in the hole. The soil around the roots is then firmed up. Proper spacing between seedlings is one of the most important factors which affect the bulb size, consequently, total yield. Close spacing of 15 cm between rows and 7–8 cm between plants has been found most conducive to obtain high yields. Plants should not be set deeper than 2–3 cm. Field experiments show that seedling size is directly related to yield, with larger seedlings giving higher yields. When overgrown seedlings are planted, they are topped, but, there is no discernible difference between topping and non topping inasmuch as yield is concerned. It is advisable to irrigate the field immediately after transplanting.

Direct Bulb Planting for Green Onions as Salad

To meet the demand for green onions for salad in early winter, when dry onions are priced very high, the crop can be raised for marketing in October. For this purpose, small-sized onion bulbs are planted in the field in August and September. This planting will make the green onions available in the market in the moths October–November.

Direct Planting Method

For this, the soil is first thoroughly pulverized. About 13–18 kg seeds/ha are sown in bands, 30 cm apart, in September–October. A light irrigation is given immediately after planting. Seeds begin to sprout, in a week's time, when subsequent irrigation is done every 10 days and when the plants are 6–8 weeks old, they may be thinned to 5 cm apart. Thinning should be done before the onion plants are as large as a lead pencil. Thinning is best done when the soil is moist. Care should be taken not to disturb roots.

Irrigation

Crop characteristics and season determine the exact time and application of irrigation. First irrigation should follow immediately after transplanting. Subsequent ones may be at 10–15 days intervals until the plants are a month old. Water shortage will adversely affect growth. Application of water after a dry spell would initiate new growth of onion bulbs, which cause bulb splitting and doubling. This would adversely affect quality and yield. Over supply of water during maturity will make the bulbs watery and reduce their keeping quality. The leaves also turn yellow.

Interculture, Weeding and Rogueing

Keeping the onion field weed free is very important, as weeds compete for plant nutrition and water. Careful hoeing, without injuring roots, and, rogueing before flowering ensures a good crop.

Nutrition

Nitrogen N deficiency in soil causes leaves to turn yellowish-green, curled, wilted and dwarfed. At the beginning of the ripening period, the tissues above the bulbs become soft in N deficient soils. Excessive N fertilizer application will result in huge upper growth at the expense of the bulbs. P deficiency leads to slow growth and maturity will be delayed, also, leaf color turns light-green and bulbs get few dried outer scales. Leaf tip burn, leaves becoming dark green and erect and the plant becoming susceptible to diseases and loss of vigor are the symptoms of K deficiency. Onions grown in peat soils exhibit abnormal Cu deficiency, turn pale yellow in color. The bulbs lack firmness and solidity. Application of 125-155 kg/ha of copper sulfate increases thickness of onion scales and changes color from pale-yellow to brilliant brown. B deficiency gives onion a characteristic stunted, distorted appearance. Leaf color varies from dark grey-green to deep blue-green; the youngest leaves develop conspicuous yellow and green mottlings. Shrunken areas appear, followed by ladder-like transverse cracks on the upper side of the basal leaves, and they become stiff and brittle. Deficiency can be corrected by applying borax at 18-20 kg/ha. Appearance of irregular elliptical-shaped areas almost white in color near the end of the leaves are the symptoms of Mg deficiency which can be corrected by applying 60–100 kg/ha magnesium sulfate.

Fertilizer Recommendations Application of 550 kg/ha of calcium ammonium nitrate and 350 kg/ha of single superphosphate is the optimum fertilizer rates for onion grown in Punjab and Haryana. Whole of the superphosphate and half of the calcium ammonium nitrate should be drilled at the time of sowing and the remaining half a month later after transplanting. Additionally, 100 kg/ha of potassium sulfate also be drilled at the sowing time, if soil is found deficient in K.

Harvest

Onion crop is ready for harvest 5 months from transplanting, for dry onion, green onion, however, is ready in 3 months. Onion should be pulled out of the soil by using a hand made device, locally called, ***khurpa***. This is a flat sickle-like equipment made out of cast iron with a wooden handle, which, when pushed into the soil can lift the onion bulbs without injury to the bulb. The bulbs should immediately be put/spread under shade for drying. They should never be left in the field during the scorching sun of April/May which will rot, if done, during curing. After curing the bulbs in the shade, their tops are removed, cut off by hand, or sheared by a knife. It is desirable to leave 2–3 cm of tops attached to the bulb. If cut too close the neck does not close well and decaying organisms will have an easy access to the bulb. There is a steady increase in the volatile sulfur until the plants reach maturity. The peak is reached when the tops begin to fall over. If the harvested onion is left in the field for a longer time, volatile sulfur content lowers.

Yield

Yields average 140–150 quintals (100 kg)/ha, and, early sowing, at the end of December increase it by about 12 quintals/ha. The large seedlings, when used, produce much higher yield than the crop sown with small seedlings. Seedling-planted crop produces about 25% more than the bulb-planted one.

Seed Production

The onion seed is raised from the bulbs produced in the previous season. True to type and shapely onions are graded out from the lot and are reserved as mother-seed bulbs. These are planted in well prepared seedbed by mid September. Sprouts are removed when bulbs are planted. Again, the planting of onion bulbs is done with the ***khurpa.***

Average-sized seed bulbs, weighing 10 quintals are required to plant a hectare of land. The after care is the same as for the main crop.

Isolation Distances

Since the crop is cross pollinated, no two different varieties of onion should be planted close to each other. There should be an isolation of 0.25 to 0.5 kilometer between two varieties meant for seed production, However, longer distances are required for foundation seed production.

The seed crop takes 6 months to mature. All umbels on a single stem do not mature simultaneously. In each umbel, central flower matures first of all and is followed subsequently by peripheral ones in turn. Hence, the harvesting usually should be completed in three rounds.

After harvest, the umbels are threshed and seed extracted. The fruit of the onion is a 3-lobed, 3-celled capsule. Each locule contains 1–2 seeds, which become black colored at maturity. Most of the seed is endosperm. One hectare of seed crop yields 875-1225 kg of onion seed.

Crop Protection

Insect Pests

1. Onion Thrip: This is a very serious pest of onion. The nymphs and adults suck plant sap from young leaves. The older leaves become silvered and assume blistered in appearance. Large number of thrips can cause severe injury to the onion crop, and, eventually kill the plants. The insects feed on the flower stems in the seed crop adversely affecting seed production. The pest can be controlled by spraying 0.1% DDT suspension or 0.2% endrin emulsion (1.25 kg DDT 50%) WP or 0.625 liter endrin, 20% EC in 625 liters of water per ha. It can also be controlled by spraying malathion (625 ml of malathion 50EC) as soon as the pest emerges.
2. Red spider mite: Red spider normally attacks the crop in dry weather. The insect is very tiny, found in clusters of varying sizes. Leaves become pale, sickly in appearance which gradually dry and fall. The mite sucks the plant sap. It can be controlled by dusting the crop with finely powdered 22.5 kg sulfur/ha. When attack is serious, several dustings would be required. The pest can also be checked by spraying parathion @290 g/1.225 liter of water. Parathion should not be sprayed to edible produce within 30 days of harvest.
3. Onion maggot: The severity is more in wet season. The maggots bore through the underground stem into the bulb. This results in rotting during storage. A spray of mercuric chloride 70 g in 90 liters of water can control the maggots. To pre-empt attack, onion fields are to be kept free of post harvest vegetative matter.

Diseases

Onion Blight The disease is common in the seed crop, caused by *Alternaria porri*. Black spots on leaves in early stages are seen by the diseased plants. Small sunken lesions with purple centers girdle the seed stalk. A 5.5:50 bordeaux mixture of copper fungicide controls the disease.

Bacterial Soft Rot This is a storage disease that can cause considerable damage to the harvested bulbs. It starts from the onion fields when the plants near maturity. Bacteria enter the neck tissue of the plant through dead leaves and the scales start rotting. The rot is slimy and yellow and smells like vinegar. It is caused by one or more organisms, like, *Erwinia carotovora.* To control the disease, the crop should be allowed to mature properly in the field before harvest. Tops should be allowed to desiccate as much as possible before cutting. Bulbs should be thoroughly cured before storage and the storage houses must have hygienic conditions and sufficient ventilation.

Marketing

1. Preparation of the Bulbs: Undersized and diseased bulbs are discarded when onions are taken to the market. The outer scales are usually rubbed off during the grading process to give onions a better appearance. The bulbs are usually taken in gunny bags, by bullock carts, or in loose stacks, to the market. There is no quality control and no grading is done at the farm. The lots are dumped in the market in the allotted place for sale. When no storage facilities are available, onions should be taken to the market right after harvest, if not they are likely to rot and lose weight. To get better market price, it is advisable that graded onions are bagged in mesh-gunny bags. Mechanical pneumatic preparation systems (machine) both for large and small onions are available. The prepared onions are automatically filled into plastic-woven bags, weighed and palletized for marketing.
2. Preparation of green onions: The green onions after harvest should be washed, and outer dirty and dried scale leaves should be removed completely and the roots trimmed to half an inch in length. Bulbs dressed this way look attractive. To facilitate loading, unloading and auctioning, the produce should be tied in small bundles. While the produce is being prepared for the market, it should be covered by moist gunny bags or kept under shade to prevent loss of weight.
3. Preparation of onion seed: There is no regular market for onion seed and, generally, the growers sell their produce through local seed depots. Therefore, it is very essential that the seed should be thoroughly cleaned by repeated winnowing before dispatching to market. It is advisable to prepare sealed packets of small

size. This will prevent adulteration by the grocers, and, will also ensure good market for the produce.

Grading: The practice of systematic grading of onions based on recognized grade standards is almost completely absent in India, growing much onion. The producers in some states, however, sort out the onions into different lots, first by removing rotten and diseased bulbs. Maharashtra which grows the maximum onion in India, sorts into three lots, takes the first two to assembling markets for sale, while the third is sold locally. But, there is no specific basis for this sort of grading, and, even the mode of sorting differs from cultivator to cultivator.

In order that the benefit of proper grading be shared equally by the producer and consumer it may be advisable to grade the produce. In this context it is worth noting the composition of onion bulbs and onion powder (Table 1).

Storage

Some Useful Suggestions

1. Only well ripened and thoroughly cured onions are fit for storage
2. Immature, soft and thick-necked bulb should be sold as soon as harvested

Table 1 Comparative properties of Onion Bulb and Onion Powder

	Onion Bulb	Onion powder
Moisture percentage	86.8	4.6
Protein percentage	1.2	10.6
Fat percentage	0.1	0.8
Carbohydrate percentage	11.6	74.1
Fiber percentage	–	8.4
Calcium percentage	0.18	0.3
Phosphorus percentage	0.05	0.29
Iron	0.7 mg/100 g	0.002%
Mineral matter (total ash percentage)		–3.5
Sodium percentage		0.04
Potassium percentage		1.0
Refractive index		1.041(optical rotation - 5 degree)*
*the bulbs and fresh herb yield 0.005% essential oil which has an acrid taste and unpleasant odor		
Vit A		175 (IU)/100 g
Vit B		0.42 IU/100 g
Vit B2		0.06 IU/100 g nicotinic acid
Vit C		14.7 mg/100 g
Nicotinic acid		0.6%
Calorific value		370 calories/100 g

3. The store should be thoroughly ventilated and should have comparatively low ambient temperature and dry humidity-free atmosphere. Ideal storage temperature should be 0 degree centigrade, 64% relative humidity, as per USA recommendations. In India, relative humidity can be between 70 and 75%.
4. A separate store room should be used for onions, if not, the odor may be absorbed by all other food stuff stored nearby
5. In India, onions are stored in commercial stores, *kupas*, and cold storages.
6. Onions are usually stored in such rooms from June – September, and, rotten onions are frequently removed
7. When onions are stored in such *kupas*, care should be taken that rooms are airy and bulbs are rolled over regularly so that the sides of the bulbs are exposed to air and light at regular intervals for proper curing
8. In rural areas, the most popular method of storing onion bulbs is in the *kupas*. After curing, the bulbs under the shade of trees or in an open corridor, are shifted to the *kupas*. It is believed the loss in onions is least when stored by this method
9. In Maharashtra, special types of store-houses are constructed for storage, called *Basha* in Nasik district and *Chalies* in Sholapur district. They have the following dimension: 1.22–1.53 m (broad) x 2.44–3.05 m (high), length varying according to the size of the onion bulk to be stored, usually not exceeding 34.08 m. Side walls are made of thin bamboo. Floor is raised to a height of 0.15–0.31 m above the ground. Bulbs are placed in Iayers 7.62–10.16 cm and rolled over periodically, and, in the process, diseased and damaged bulbs are culled out. Capacity of store houses is about 20,000 kg. and storage usually extends from May to September.
10. Total losses in storage is estimated to be about 25% due to drying, sprouting, rotting, and attack by rats.

Cold Storage

The onions are also kept in cold storage during summer. The best temperature to store onion is 1–2 degree centigrade with relative humidity of 70–75%.

Inhibitors against sprouting during storage:

Inhibition of sprouting during storage of onions is important and many growth inhibitors have been tried. Best results are obtained by spraying maleic hydrazide 2000 ppm 1 week before harvest.

Irradiation

The weight loss, sprouting and rooting of “Co 44”onion variety were the least in bulbs treated with gamma rays and stored in perforated trays.

Effect of Nitrogen Fertilization on Storage Behavior

Storage behavior of onion bulbs grown with graded rates of N (0, – control – 40, 80, 120, 160 and 200 kg/ha) indicated significant differences in bulb firmness, total soluble salts, and sugars. N significantly decreased dry matter, increased average bulb weight and total yield, but, there was no change in weight loss during storage. Increasing N levels led to increased rotting of bulbs. Increase beyond 160 kg N/ha resulted in steep increase in storage loss. The variety Arka Niketan showed significantly lower storage losses than the variety Arka Kalyan, which could be stored only for 3 months, whereas Arka Niketan could be stored upto 6 months.

Curing

Onion bulbs immediately after harvest are allowed to lose extra moisture by keeping them in field under shade. It is necessary to have thin and closed necks and firm outer skin. The period of curing may be from 2–3 days up to 1 week, depending on the season. Curing in sun should be avoided as it causes heavy spoilage ad weight loss.

Neck Cutting

Leaf tops are removed after initial curing; 2 cm neck portion is left with the bulbs to obtain complete closure and thin neck. To avoid injury to bulbs, sharp implement is used. Final curing is again followed up after neck-cutting for the required period.

Sorting and Grading

To avoid loss due to rot in transit, storage and marketing, and to maximize profit for the farmer, diseased thick-necked, damaged and bolted bulbs are sorted out and lots are then graded into different categories based on bulb size. Medium-sized bulbs (about 40–45 mm in diameter) are preferred, in general, and hence, fetch attractive price as compared to extra big bulbs (over 60 mm in diameter) and small-sized bulbs of 20 mm diameter.

Packaging

Type of packing depends on the transportation distance and consumer choice. For local market, onions are generally transported unpacked, often using gunny bags to put the bulbs into. Each bag contains about 80 kg. But, caution must be exercised to minimize storage time as increased time in gunny bags will lead to bulb rotting. When packing for distant market jute bags (Hessian bags) are preferred. These have 8 × 8 mesh per square inch, weighing about 200–250 g. A 40 kg onion packing is commonly used for internal marketing, whereas 10–25 kg onion packet is used for export.

For consumer packing, normal discarded magazine paper is used for 1–5 kg onion supply. In overseas markets 10–25 kg jute bags (open mesh) or nylon-netted bags are used. Some consumer packs of 2-5 kg are also made of paper mesh or kraft paper.

Filling and closing of bags: Filling and closing operations are undertaken manually after proper sorting and grading. To weigh, 50–100 kg platform balances and also counter balances are used.

Labelling: For marketing within the country no labeling is required. But, for overseas market, labeling must be done. The name of the exporter, weight, destination and importers name are specified on bags using stencils. Agmark (Indian) labels specifying variety and size are put as "Quality Certificate".

Effect of Packing Material on Quality

Since onion bulbs are live material air is needed for respiration. Thickness and aeration offered by packing material affect quality. Experimentally it has been shown that losses in transit are least when onions are packed in hessian bags, weighing 260 kg each with 9 × 10 mesh/square inch.

Transportation: Effective transportation is a prerequisite for economic gains both for the onion farmer and the customer to minimize transportation losses since onion is perishable. The mode of transportation varies with the quantity to be transported and distance to be covered.

Road transport: Since quick transportation is required because of the perishable nature of onions, road transport is preferable for short distances. Transportation data in India shows that between Nasik (Maharashtra state) and Karnal (Haryana state), covering a distance of 1800 km, only 1.5% transit loss is seen. Road transport thus provides the following advantages:

1. Quick transport
2. Door-to-door delivery
3. Cheaper for comparatively short distances
4. Suitable for small quantity at a given time
5. Lower transit losses

Rail Transport Rail transport is meant for heavy loads and long distances. However, the main disadvantage is the ready availability of wagons to carry the load. More often than not, they are unavailable on demand, and, one has to wait for long during which the onions would rot. Absence of proper storage conditions is another drawback. Main advantage is cost of transport and security of produce, since, theft can occur during road transport. Rail transportation experiments conducted in India between Nasik (Maharashtra State) and Chennai (Tamil Nadu) and Nasik and Hyderabad (Andhra Pradesh) showed that physiological losses in ordinary wagons varied from 3–31%. Use of modified wagons which have proper ventilation, suitable cushioning at the base, with the help of paddy husk, straw etc., and side-wall lining with bamboo mats largely minimize transportation losses.

Sea Transport Onions are shipped from the nearest sea port to the production center. Maharashtra state produces the maximum onions and Mumbai is an ideal sea port for shipping onions. Onion bags are stacked in 6–7 tiers or even 14–17 tiers. Often onion is also carried along with other cargo. Proper ventilation is a must during sea transport. No systematic investigations have been made on the merits of sea transport. Quite a few times, the entire shipment is spoiled due to various factors, improper ventilation, improper packing and so on.

Pre-requisites of Good Onion Storage Effective storage can only be done when the pre-requisites of storage and factors affecting it are properly understood. Onions can ordinarily be stored during 4–5 months in sufficiently ventilated places. The particulars of storage requirements in onions are as follows:

1. Suitable varieties: Onion varieties grow in *Rabi* (winter sowing summer harvest) are suitable for storage owing to their good keeping quality, whereas *Kharif* (summer sown winter harvest) onions are poor storers because of their low dry matter contents, low refractive index, ad a high rate of water loss. Varieties Pusa Red, Nasik Red, N 2–4-1, Agrifound Light Red and Arka Niketan have good keeping quality. White colored varieties grown either in *Kharif* or *Rabi* seasons are unsuitable for long storage because of their poor keeping quality.
2. Proper Drying and Curing: Bulbs should be properly dried and cured so that excess moisture from the bulbs is removed and the outer scales develop uniform color. Poor curing results in heavy losses due to drying and microbial spoilage during storage. Well cured and dried onions have firm outer skin which gives them thin and dry appearance. After curing, bulbs should be conditioned before storage.
3. Good Bulb Quality: Bulbs meant for storage rare required to have high total solids (TS). Bulbs with high TS values generally have a feel of hardness and heaviness. They should have closed necks preferably about 2 cm long and firmly attached to outer scales. Length and number of roots should be minimum and should be free of attached soil particles. Thick necked, bolted, doubled and diseased bulbs should be sorted out before storage.

4. Careful Handling: Onions need careful handling to prevent damage due to physical bruises and losses of the outer protective scales. They should not be dropped down from more than 3–6 feet height. However, onions can be dropped on other onions already on the storage floor due to the cushioning effect. Onions should be stored in stacks than in bags. The height of the stack may be as high as 7–10 feet. Storage losses will be less when the stacks are 4–5 feet high.

Factors Affecting Efficient Storage Different factors, including environmental ones, greatly influence the effective storage of onions, thereby, affecting extent of damage during the storage period. The following ae the important ones:

1. Cultural Practices: The storability of onions depends much on the different cultural practices, such as, fertilizer use, irrigation, and so on. Excessive N fertilizer use produce large-sized thick-necked bulbs resulting in increased losses due to sprouting and rotting. Similarly, late application of N fertilizer produces bulbs of poor keeping quality. The P and K fertilizers have little, if any, influence on storage. Irregular irrigation, specially prolonged interval between two consecutive irrigations leads to splitting of bulbs. Prior to a fortnight to harvest, onions should not be irrigated. Harvesting 1 week after 50% top fall is reported to have recorded minimum storage loss.
2. Plant protection measures in the field: Inadequate and ineffective plant protection measures lead to carry over of microorganisms, which cause storage problems. Hence, it is very important to control both disease onset and insect pest attack in the field itself. Black mold, *Colletotrichum* blight, basal and stalk rots, onion fly and thrips are important diseases and insect pests of onion which are carried over from field to storehouses, if not effectively controlled in the field itself. Affected bulbs have poor luster, discoloration and softness. These bulbs also trigger spread soft rot bacteria during transit and storage.
3. Temperature and Relative Humidity: Sprouting is affected by storage temperature ad rooting by relative humidity (RH). Onion keeps best at 0 degree centigrade and 64% RH. At 0 and 30 degree centigrade sprouting was lowest, but, at 15 degree centigrade it was highly pronounced. Higher RH stimulates both sprouting and rooting and higher temperatures remove moisture from onion bulbs due to activation of physiological processes, including, respiration and transpiration. Increased moisture around onion bulbs favors rotting caused by microorganisms. There will be increased sprouting and rooting when it rains, and, when climate is hot, rotting of bulbs is pronounced. The bulbs prior to storage should be pre-conditioned to favorable temperature inside the storage house to minimize microbial spoilage.

   Types of storage structures: Excepting for capacity, floor type, material used for sides and roof, all onion growing states use similar storage houses. In northern India bamboo nets are used for sides and roof is of thatch covered by jute cloth. The size is about 1.2 m × 1.2 m × 3.0 m storing about 4000 kg bulbs. Bottom net is fixed at 15–20 cm height above ground level for efficient aeration.

Physiological and Mineral Losses During Onion Storage Experiments conducted in Model 2-tier and Model-3 tier storage houses demonstrate reduction of total losses as compared conventional godowns. Losses due to drying (22–24%), sprouting (4–5%),and, microbial damage (5–10%) have been noted. Average of 3 years overall losses were 27.7% in the above-mentioned storage houses compared to 55.6% in conventional types, during 5 months storage at ambient temperature.

Microbes Causing Storage Rots in Onions The microorganisms causing storage rots include bacteria *Erwinia, Pseudomonas* and fungi *Fusarium oxysporum, Sclerotium rolfsii, Aspergillus niger, Rhizopus* sp. etc. Among insect pests, onion fly *Hylemyia antiqua* and thrips invade onions causing substantial damage during storage.

Ways to Reduce Storage Losses in Onions Storage losses in onions can be reduced by the following ways:

1. By checking spouting during storage achieved by field application of maleic hydrazide, 2000–3000 ppm 75–90 days after transplanting
2. By maintaining sanitation in and around store-house using disinfectants such as formalin@0.65%
3. By treatment of bulbs soon after harvest with 0.1% bavistin solution

Packaging and Storing Onion in Jute Bags Packaging onions in very thin jute bags, 260 g in weight (9x10 mesh/square inch) is shown to reduce losses in comparison with other packing materials. This confirms the usefulness of jute bags for large scale packing.

Mechanism and Extent of Water Vapor Losses from Onion During Storage When onions are stored at ambient temperature (>20 degree Celsius) and low RH, the extent of sprouting is low because of the high temperature which maintains dormancy. However, the losses due to desication are high, reaching up to 30% during storage of 5 months. When the outer scale of the bulb dries up due to water loss, the water from the inner scales evaporates. Loss of water vapor occurs mainly through the neck region of the bulb. Immature bulbs lose more weight and bulbs with thicker neck allow greater evaporation of water from interior. Onions stored at 21–35 degree centigrade and ambient RH 2% showed maximum loss of water vapor when both neck and base were blocked, whereas onion stored at 21–35 degree centigrade and ambient RH 50–80% the losses showed reverse trend during 45 days of storage. Contrary to the established belief that loss of water vapor occurs mainly through neck region of the bulbs, it was found that it was largely through sides, although loss of water vapor does occur through neck and base regions also. Of the total loss that occurred in the first 3 days, 41% was from the sides, which increased to over 90% during 45 days of storage under both conditions.

Value Added Products from Onion A number of value added products are being manufactured on a commercial scale from white onions, wherein dehydration is involved. For instance, dehydrated onion, onion flakes, onion rings, kibbled onions, onion powder and onion salt are some of these. Small pearl onions are used "whole" in vinegar pickle manufacture. Of late, onion paste is also being manufactured which is increasingly popular in curry making.

Quality Parameters of White Onion for Dehydration Purposes The following are the quality parameters of White onion to make onion powder:

1. White colored flesh
2. Full globe to tall global shape of bulbs of 5–6 cm diameter
3. High total solid content above 15% preferably 20%
4. High degree of pungency
5. High yield
6. Good keeping quality – at least 2–3 months
7. Freedom from joints
8. Disease free

Dehydration of Onion Dehydrated onion is produced by removing water from raw onions to a maximum level of 4.25% and then milling it to a specific particle-size. Onions are dehydrated without blanching or sulfiting in order to protect the enzyme system which develops onion flavor when onion cells are cut or broken. When the cells are broken, the onion enzyme allinase is free to contact certain onion compounds which react to form numerous volatiles responsible for the characteristic onion flavor.

Prior to drying, onion is cleaned and peeled, roots and tops removed, the peeled onions are washed and sliced. It is important to note that while blanching cannot be used to control bacteria, the thorough cleaning operation and the conditions of drying remove a large number of microorganisms. Those remaining are primarily of the spore-forming types which are resistant to time-temperature treatment of drying. None of these have been found to have any health hazard of significance.

Onions are dried on a stainless steel continuous conveyor belt, which passes them through 3 or 4 stages, following a carefully controlled time/temperature program. The optimum amount of moisture is removed at each stage, the onions continue on to each successive belt maintained in a condition favoring circulation of heated air through the product. The conditions can be controlled to deliver onions with a maximum of 5% moisture, without any heat damage to the product.

Onion Powder Onion powder is prepared by grinding dehydrated onion slices in a hammer mill to a suitable mesh. It is highly hygroscopic and hence, the important precaution regarding its storage is to keep it in air-tight containers in a cool, dark and dry place, failing which, it may absorb moisture, become granular, caky, pasty and ultimately get mold attack.

Composition (Table 2)

Onion Salt It is prepared by mixing 19–20% of onion powder with 78% free-flowing pulverized, refined, iodized table salt and 1–2% anti-caking agent like any-hydrous sodium sulfate which prevents water absorption and caking of the onion powder during storage.

International Quality Standards for Dehydrated Onion and Onion Powder The international (ISO) quality standards both for the dehydrated onion and onion powder are summarized in the following table (Table 3):

Uses On account of the presence of several sulfur compounds, onion has anti septic properties. Onion has stimulant, diuretic and expectorant properties and is considered useful to control flatulence and dysentery. Freshly prepared onion juice has moderate bactericidal properties. Onions are used as salad, and are cooked in several ways in all types of the famous Indian curries. They are baked, fried and used in fresh or dehydrated powder form in soups, pickles, sauces etc. Thy are also eaten raw as salad which helps in lowering blood pressure and blood cholesterol.

Table 2 Composition of Onion Powder

Property	Content
Moisture	4.6%
Protein	10.16%
Fat	0.8%
Fiber	8.40%
Carbohydrate	74.1%
Mineral matter (Total ash)	3.5%
Calcium	0.3%
Phosphorus	0.29%
Sodium	0.04%
Potassium	1.00%
Iron	0.002%
Vitamin A	175 IU/100 g
Vit B	0.042%
Vit B2	0.6%
Nicotinic acid	0.06%
Vit C	14.7 mg/100 g
Calorific value	370 calories/100 g

Table 3 International (ISO) Physico-chemical Quality Specifications for Dehydrated Onion and Onion Powder

Physico-chemical characteristics	Dried slices flakes, pieces, kibbled onions	Onion powder
Moisture content % (m/m) maximum	8.0	6.0
Total ash content% (m/m) dry basis maximum	5.5	5.5
Acid soluble ash% (m/m) (dry basis) maximum	0.5	0.5
Extraneous matter(rots,papery skin etc.)	0.5	0.5

Source: ISO 5559-1983 International Standard for dried onion and onion powder

Reference

Chaudhury B (1983) Vegetables 7th Reprint Edn. National Book Trust, New Delhi

Garlic

Botanical name: *Allium sativum* Linn: family: Liliaceae

History and Origin Garlic (local name *Lassan*), belonging to the onion family, has long been recognized all over the world as a valuable condiment for making many foods and as a popular folk remedy or medicine for various ailments and physiological disorders. More than three decades ago it attracted much global attention as an important spice in controlling cardiac diseases, and, continues to be so, even now. Records show that Egyptians regularly consumed garlic 5000–6000 years ago, that the Greeks loved it, while the Romans of Julius Caesar's days, did not, but, the modern Italians love it.

Garlic is a native of Western Asia and Mediterranean area. It has been in cultivation since centuries. Europeans, especially the Italian and Spaniards, have regularly used garlic during more than 2000 years. The Spaniards are believed to have brought garlic to the New World (USA), where it became an immediate favorite with the Red Indians. They liked it better than any other root or bulb crop imported from Europe. A teasing aroma of garlic drifts over the American cuisine. During the last decade there has been a 1000% increase in garlic use in the USA. Credit for the zooming popularity of garlic must go to the convenient dehydrated garlic products, such as, instant minced garlic, instant garlic powder, granulated garlic and garlic salt. The first three forms are pure garlic, while garlic salt is garlic powder combined with 75% table salt and 1.2% desiccant. Instant minced garlic is ideal in dishes which require cooking or in which garlic bits can stand until flavor is released. All other forms release flavor instantly.

India has been the largest producer, consumer and exporter of garlic during the mid eighties. It produces about 2.2 lakh tonnes of garlic from about 70,000 ha. In Tamil Nadu garlic cultivation is confined to the hilly tracts of Nilgiris District.

Garlic is a hardy, bulbous, rooted perennial plant with narrow leaves, and bears small white bulbs. The compound bulb consists of 6–34 smaller bulblets called "cloves", which are surrounded by a common thin white or pinkish papery sheath,

K. P. Nair, *Minor Spices and Condiments*,
https://doi.org/10.1007/978-3-030-82246-0_3

skin or membrane. Garlic has a stronger flavor than onion or its other allies. Recovery of cloves is 86–96% on average 93.7%.

The plant is cultivated in temperate as well as tropical climate. It is grown from bulblets or cloves as a mixed crop or monocrop. It requires a fertile soil. Garlic has since long been cultivated throughout India as an important minor spice or condiment crop. During 1984–1985 the acreage under garlic, as an irrigated crop, was about 95,000 ha in India. Since then, the area has substantially increased. The important states growing garlic are Tamil Nadu, Gujarat, Maharashtra, Karnataka, Andhra Pradesh, Madhya Pradesh, Orissa Uttar Pradesh and Haryana.

Area and Production Globally, garlic is grown in 380,000 ha, as per FAO statistics, with a total production of 231,500 tonnes, with an average yield of 6.09 tonne/ha. India is the most important garlic producer in the world. In India, in 1994–1995, area wise, of the total 959,000 ha, it is grown in the following sequential order, Andhra Pradesh, Bihar, Gujarat, Haryana, Jammu and Kashmir, Karnataka, Madhya Pradesh, Maharashtra, Orissa, Punjab, Rajasthan, Tamil Nadu and Uttar Pradesh states. Madhya Pradesh produces the largest, followed by Gujarat, Orissa, Rajasthan, Uttar Pradesh, and Maharashtra (both states have the same acreage). Production wise, in 1994–1995, Madhya Pradesh tops (1,199,000 tonnes) with Andhra Pradesh producing the least (11,000 tonnes).

Export Garlic and garlic products, especially garlic powder and dehydrated garlic are the most important export items. Among the overseas countries, England tops among the importing countries, while Japan imports the least (1992–1993), inasmuch as dehydrated garlic is concerned. With regard to dehydrated garlic flakes, Japan tops while Maldives imports the least (1992–1993).

Varieties No distinct garlic varieties are recognized by the trade in northern India, but, two distinct types, namely, *Fawari* and *Rajalle Gaddi*, with slightly bigger bulbs are grown in the Bellary district of Karnataka State, in Southern India. Chowdhury has divided the Indian varieties into two groups as follows:

1. Small and medium-sized bulbs with small to large number of cloves
2. Big-sized bulbs and cloves

   The second group is generally cultivated in the hilly regions of India and also in Gujarat State.

Quality Parameters for Garlic Bulbs for Breeding/Production Breeding for production of new varieties with large white bulbs, uniform in shape and size, bold and compact cloves, high yield, and, resistance to diseases and insect pests, has been carried out in the Punjab Agricultural University, Ludhiana, India and also in Gujarat State.

During the course of the development of a patented process to manufacture garlic powder from Indian garlic at the Central Food Technological Research Institute in Mysore, Karnataka State, garlic grown in important States in India were tested and it was observed that Jamnagar (Gujarat State) garlic was the biggest and the best, giving the maximum recovery of dehydrated peeled garlic and garlic powder of good pungency and anti-bacterial activity (Pruthi *et al* 1980).

Crop Management

Soil and Climate Garlic grows best in the same type of soils which are suitable for onion, as well, except that it favors a very fertile soil and grows best at higher elevations (900–1200 above mean sea level). A well drained, moderately clayey-loam soil is best suited for garlic cultivation. It requires a cool and moist period during growth and a relatively dry period during crop maturity. It takes longer (4–5 months) than onion to mature and is consequently grown as a late-season irrigated crop. In South India it is grown in rotation with *Ragi* (*Eleusine coracana*), chillies, maize, potato and beans.

In India, mainly small garlic bulbs with more than 20 cloves are cultivated. The export specifies bulbs of more than 40 mm in size. Recently bigger cloved garlic varieties like G-282 for plains and Agrifound Parvati for Northern Indian hills are are recommended to be grown mainly for export. Varieties with small cloved types are Agrifound White, Yamuna Safed.

Garlic bulbs 40–60 mm in diameter with 10–15 cloves in each bulb are preferred for export. China, Spain and France are the major suppliers of garlic having bigger cloves of the above-mentioned specifications. Wholesale price of garlic varies with the locations where grown and seasons.

Manurial Requirements Since garlic requires a more fertile soil than onion, the most important mineral nutrient for garlic is nitrogen. About 125–150 kg of nitrogen/ha should be applied in two split doses to get the best results.

Seed Rate and Sowing Methodology Garlic is propagated by cloves which are first detached individually from the bulb. About 350–500 kg of cloves are required to plant a hectare. To get the maximum yield, the cloves are dibbled at a spacing of 15 cm between the rows and 7.5 cm between the cloves within the rows. Thc field is divided into small plots to facilitate proper irrigation. The crop is sown from September to November in the plains in northern India. The crop is ready for harvest when the tops turn yellowish or light brownish, and show signs of drying up, usually about a month or so after the emergence of seed stalks. The bulbs are lifted, freed from the sticking soil, and the leaves are tied up at the top. The bulbs are cured during 3–4 days in the shade before storing them in an ordinary room. Tops are removed before marketing the produce. Thoroughly cured garlic store well in ordinary well ventilated rooms/store.

Pests and Diseases of Garlic They are similar to those of Onion (please refer the second chapter on Onion).

Yield After removing the tops, the yield per hectare may vary from 4000 kg to 10,000 kg/ha, depending on the soil, variety and care bestowed on the crop.

Tissue Culture of Garlic A number of researchers have investigated the prospects of tissue culturing in Garlic. A comprehensive review (Priyadarshan et al. 1989) details this aspect.

Marketing

Smoking In hilly areas of Tamil Nadu, where garlic is widely grown, harvested garlic bulbs are subjected to a smoking process, which enhances the keeping quality of the bulbs. This smoking process is inexpensive, as it is carried out with ordinary ovens in farmers' houses.

Packaging The cured bulbs are packed in gunny bags as and when they are ready for dispatch for sale. About 50% of the produce reaches the main assembling market during February-April. The size of the garlic bag varies. For instance, in Andhra Pradesh and Karnataka it is 0f 90 kg capacity, while in other states it can vary from 40–60 kg.

Grading Export of dehydrated garlic and garlic powder are subjected to Compulsory Quality Control and pre-shipment inspection administered by the Directorate of Marketing and Inspection, Government of India in Faridabad/Delhi. According to AGMARK Grading Rules (1984), separate schedules for 4 grade designations (big, medium, small and mixed) of following types/varieties of garlic have been prescribed:

1. Pune/Nasik Garlic
2. Madurai Plain Garlic
3. Madurai Hill Garlic (smoked/unsmoked)
4. Jamnagar (Gujarat) Garlic

They have been later described along with the garlic powder etc. The grade classification is based on the diameter of the bulbs, namely, Big Bulb: minimum 35 mm, Medium Bulb: 25 mm, Small Bulb: 12 mm. Several physical quality parameters have been prescribed (DMI 1968). These are then filled into gunny bags of jute cloth, which is fairly thin and of see-through grade. The gunny cloth is technically described as hessian cloth of specification 6 oz./yard or width of 39 inches. Each bag can accommodate 50 kg of garlic; 300–350 such bags are loaded onto a truck and transported to wholesale merchants. The product sold in Hyderabad market is received from Madhya Pradesh, Uttar Pradesh and Gujarat (Anwar 1988).

Curing, Packaging and Storing

Curing

The bulbs should be cured for 3–4 days in shade before storing them in an ordinary room (Pruthi 1987). Bulbs can be dried in sun even for a week. The garlic bulbs are strung out or hung on the horizontal bamboo poles fixed in ordinary well-ventilated rooms, where it is stored fairly well until the market price is favorable. They are then removed from the poles and the tops are cut off and the bulbs are packed in gunny bags to be sent to the market. 80% of the farmers use the gunny bags for storage while 16% spread on the floor and just 4% in well ventilated wooden structure.

Packaging Garlic (whole) is packed in gunny bags without any inner lining. Size or weight of bags varies depending on the assembling/terminal markets. For instance, in Andhra Pradesh and Karnataka, garlic is packed in 90 kg bags, while in other states the farmers use 40 kg or 60 kg bags.

The Garlic Grading and Marking Rules provide for packings of 18, 25, and 50 kg. Only sound, clean and dry containers made of jute, wooden crates, netting, bamboo baskets, or palm leaf baskets should be used for packing garlic. The containers should be free from any insect infestation or fungus contamination and should be free from any undesirable smell. Each container should contain garlic of any one grade designation only. Detailed investigations have been carried out on packing and storage of garlic powder in different containers (Singh et al. 1959).

Storage Garlic (whole) packed in gunny bags should be stored on wooden dunnage in warehouses of Central Warehousing Corporations (Pruthi et al. 1959).

Experiments were carried out to investigate effect of storage temperature of garlic cloves before plating on quality, plant growth and bulb development in the field. Sprouting and rooting were accelerated before winter by low (5 degree centigrade) and room (15–20 degree centigrade) storage temperature, while both were inhibited by high storage temperature (25–30 degree centigrade). By storing garlic at low temperature, not only the percentage of secondary growth and bolting rate was found to increase, but, the date of clove differentiation was enhanced. Room temperature was more beneficial for commercial yield than any other storage temperature. When garlic was stored in a polythene film bag, more cloves decayed during storage, compared to that stored in a net sack. Effect of low temperature treatment on garlic plant growth was enhanced in northern ecotype *Seosan* compared with southern ecotype *Cheju,* and low temperature treatment was more effective on growth and bulb development of garlic than the high room temperature treatment (Bottcher and Gunther 1994).

Effect of N, P, and K on Loss of Weight in Garlic Bulbs During Storage Three levels of N (0, 100, and 200 kg/ha), P (0, 50 and 100 kg/ha) and K (0, 50 and 100 kg/ha) were tested at the Agricultural College, Dharwad, Karnataka State on their effects on weight loss in garlic. Spacing adopted was 15 cm × 7.5 cm. Urea, single super phosphate and potassium sulfate were, respectively, the carriers of these elements. Half the dose of N and full of P and K were applied at sowing and the other half of N 30 days later. In general, per cent loss in bulbs was more where N was applied and weight loss was reduced by both P and K application, but, the effect was significant in the case of K application. Rate of weight loss was much higher (7.72%) in the initial period of storage and during later 150 days after harvest, it was minimum- 2.3% (Setty and Sulikeri 1990).

Physiological Disorders in Garlic Bulbs Following disorders are reported:

1. "Rubbering" before harvest (Peter et al. 1995): Observed in fresh garlic after harvest. It can be identified by dent formation with the pressure applied through the tip of the thumb. The dent returns to original form by the release of the pressure. These rubberized bulbs become spongy when dried. This is a worrisome problem in Tamil Nadu hills growing garlic
2. Premature sprouting of the bulbs: Cloves, after initiation, continue to sprout in the field by producing leaves instead of bulbing. The sprouts from the cloves will burst out through leaf sheath. Sprouting leads to splitting of cloves which reduces market quality of garlic.

Reasons for Rubbering and Premature Sprouting The following reasons could be attributed to rubbering and premature sprouting:

1. Both are noticed in low lying areas of watershed where there is heavy deposition of nutrients along with flowing silt following heavy rain
2. Fields receive excessive irrigation
3. Higher application of nitrogen fertilizer
4. Following increased formation of rubberized bulbs due to excessive N fertilizer application, increased infestation by Thrips also is seen during later stages of the crop. The thrips lacerate the leaves and cause damage to the crop. Through lacerated wounds formed by the thrips, blast disease follows, caused by *Botrytis allii*, when the crop is about 3 months old, which further causes deterioration of the bulbs.
5. Short duration garlic crop (Mettupalayam type) is more prone than the long duration one (Singapore type).
6. Premature sprouting of bulbs is more during the main crop (April-August) when there is excessive moisture in soil at maturity of the crop due to heavy rains
7. Delayed harvest during rainy season increases the chance of premature sprouting and splitting of the bulbs
8. Wider spacing of cloves at the time of planting increases uptake of N and water by the individual plants which enhances rubbering of the bulbs and premature sprouting (Peter et al. 1995)

Control Measures for Rubbering, Sprouting and Bulb Splitting Disorders The following control measures must be considered:

1. Currently, garlic farmers are habituated to apply heavy doses of chemical fertilizers plus organic ones, for instance, they apply vegetable mixtures at the rate

of 2500–2700 kg/ha. But, application of N, P2 (2 substrate please) O5 (5 substrate please) and K2O(2 substrate please) at 75 kg/ha and 50 tonnes/ha of well-rotten farm yard manure (fym) plus 50 kg/ha of magnesium sulfate combination gives the highest yield of healthy bulbs without rubbering or sprouting phenomena.

2. When crops like potato, cabbage and carrots precede, which are heavily fertilized, and garlic follows in the same field, it is always advisable to apply chemical fertilizer at smaller rates based on precise soil testing. In this connection, a revolutionary soil management technique, for precise soil testing and fertilizer application, now globally known as "**The Nutrient Buffer Power Concept**" detailed in the book Intelligent Soil Management For Sustainable Agriculture – The Nutrient Buffer Power Concept (Nair 2019, Springer) would come in handy
3. Inasmuch as N fertilizer is concerned, it is preferable to apply Ammonium sulfate, not urea, because the former carries sulfur in addition to nitrogen, and, sulfur enhances pungency of garlic
4. Application of N and K fertilizers in splits, after a basal dose of 50%, one at 45 days after sowing and second at 60 days after sowing, prevents rubbering and sprouting/splitting of bulbs
5. In soils with high N level, when K level increases, particularly beyond 75 kg/ha up to 150 kg/kg as K20(2 subscript please), premature sprouting and rubbering diminishes. There is significant interaction between N and K, indicating that deleterious effects of excess N could be alleviated by increased application of K application
6. Application of burnt ash (a good source of K) a month after planting reduces both rubbering and sprouting
7. By closer planting of cloves at 15 cm × 7 cm (between rows and between plants) healthy bulbs can be harvested
8. By spraying micronutrient mixture containing boron at 0.1%, and sodium molybdate at 0.05%, besides 18% excess yield, healthy bulbs without premature sprouting and rubbering can be harvested
9. Spraying of growth regulators, such as Cycocel, or maleic hydrazide at 1500 ppm (1.5 g in 1 liter water) reduces premature sprouting and rubbering by 80–90% and results in 15% increased bulb yield (Peter et al. 1995).
10. Pre- Harvest Treatment to Prolong Post Harvest Storage life of Garlic Bulbs: Pre-harvest maleic hydrazide (MH) sprays have been used to prolong post harvest storage life of garlic bulbs. Spraying of 3000 ppm MH 3 weeks before harvest and 2000 ppm 2 weeks before harvest helps in this effort. These sprays inhibit sprouting for 300 days and reduce loss of weight of bulbs during storage. No appreciable adverse effect was observed on yield.

Post-Harvest Treatment for Extension of Storage Life of Garlic Bulbs The storage life of garlic bulbs can also be extended when small-sized bulbs are selected for storage and packed singly in polythene bags and stored at 0 degree centigrade to 1.6 degree centigrade and irradiated with 6 Krad of cobalt-60 gamma rays. There is storage loss of 12.2% over 150 days when garlic is stored at 1.5 degree centigrade at RH 75%, as compared to 42.4% loss in storage at ambient temperature. The UV light treatment for 30 minutes in cold storage further reduced the loss to 8.6% (Anwar 1988).

Preparation of Garlic Bulbs for Automatic Peeling and Processing The bulbs are beaten with a wooden stick to separate bulblets. Food grade edible oil is applied to bulblets manually and then these are put in small gunny bags and rubbed to remove the outer skin of the bulblets. The material is then winnowed to separate skinned bulblets which are used to manufacture garlic paste or for dehydration to prepare dehydrated garlic.

Automatic Peelers of Garlic Cloves Cloves are placed on a shake table and spread into monolayer, while being moved forward. Cloves drop onto a slopy feed tray and are positioned individually in cups (via flexible cylinders) into which a nozzle extends to provide a timed blast of pressurized air for peeling. Skins are removed under the action of a suction fan.

American Patent on Garlic Peeling Motor-driven patented equipment for automatic peeling of garlic or similar agricultural products is described below:

It has a housing, forming a processing chamber and an abrasive rotary disc in the chamber which abrases the skin or husk of the product. A moveable gate partially obstructs a discharge slot at the side of the chamber, but, in its lowered position, the gate permits continuous ejection of peelings while retaining the product. At the end of each peeling operation, the gate is raised enabling centrifugal ejection of the product by the rotary motion of the disc. A fast, high throughput is obtained because manual unloading and separation are not required. The equipment may be an accessory which engages on a food processor base (Garcia Lopez 1993).

Taiwan Patent Peeling and Drying Garlic Garlic drying and patented peeling equipment is described below:

Raw bulbs are stored in an upper tank connected to a hollow-walled drying funnel into which bulbs can be dropped. Bulbs are dried by hot air, passed through the inner surface of this funnel (from a hot air tube connected to the outer wall), and are mechanically pushed down into a membrane peeling cylinder. This operates through a helical current of compressed air provided through tapered nozzles and has an exhaust tube into which membrane pieces are guided by a curved plate.

Storage Pests Stored garlic is infested with larvae of *Ephestia elutella* which eat bulblets from inside. Fumigating bulbs with phosphine at 1–4 tablets/cubic meter for 12 h controls storage pests without adversely affecting clove sprouting. Treatment may be repeated every 2 months (Cosenza et al. 1981).

Composition

Fresh Garlic (Table 1).

Dehydrated Garlic (Table 2).

Table 1 Composition of fresh garlic

Property	Content
Moisture	62.8%
Protein	6.3%
Fat	0.1%
Mineral matter	1.0%
Fiber	0.8%
Carbohydrates	29.0%
Calcium	0.03%
Phosphorus	0.31%
Iron	0.0001%
Calorific value (food energy)	1.42 calories /100 g
Vit A	0
Nicotinic acid	0.4 mg/100 g
Vit C	13 mg/100 g

Table 2 Composition of dehydrated garlic

Moisture	5.2%
Protein	17.5%
Fat	0.6%
Mineral matter	3.2%
Fiber	1.9%
Carbohydrates	71.4%
Calcium	0.1%
Phosphorus	0.42%
Iron	0.004%
Sodium	0.01%
Potassium	1.1%
Vit A	175 IU/100 mg
Vit B	0.68
	0.08
Niacin	0.7%
Vit C	12 mg/100 g
Calorific value (food energy)	380 calories/100 g

Garlic Powder Garlic powder is one of the fastest growing value added products in the garlic industry. Not just the powder but, the Instant Minced Garlic and Garlic Flakes. Fresh garlic repels many American and European home makers because of is clinging odor, but, garlic powder is easy to use and has no "undesirable traits", as some people see it. A number of published papers dwell on the R & D of garlic (Pruthi 1994).

Commercial Forms of Dehydrated Garlic The various forms of dehydrated garlic are all produced by slicing peeled sound garlic cloves into flat slices(of thickness agreed between interested parties) which are dehydrated, graded and further processed as per requirements.

The following broad categories are recognized in the garlic trade, though commercial contracts may include modified requirements for particle size

1. Dehydrated garlic Slices or Rings: Dehydrated garlic in pieces larger than 4 mm (largest dimension) – slices and pieces of garlic are obtained by cutting garlic cloves into slices and removing broken pieces, smaller than 4 mm by sieving
2. Garlic Flakes or garlic Pieces: Dehydrated garlic passing through a 4 mm sieve, but retained on a sieve of 1.25 mm aperture- a product comprising particles without definite shape
3. Garlic Gritts: Dehydrated garlic passing through a sieve a 1.25 mm sieve but retained on a 2.5 mm sieve
4. Garlic powder: Dehydrated garlic passing through 0.25 mm sieve, a uniform product of which 95% passes through the sieve
5. Garlic salt: Comprising 20% garlic powder, 78% table salt and 2% desiccant
6. Odorless garlic powder: A product made through the patent of CFTRI, Mysore
7. Oil of Garlic: Essential oil of garlic obtained by steam distillation
8. Garlic oleoresin: Manufactured by suitable extraction technique

Patented Manufacture of Garlic Powder Pruthi (1980) describes a patent for the manufacture of garlic powder perfected at the CFTRI, Mysore, which apart from being very economical, yields a powder of better color, flavor, pharmaceutical properties, in addition to anti microbial quality. It has the following advantages:

1. About 20% of the fresh garlic produce is wasted due to respiration, transpiration and microbial spoilage on account of lack of adequate storage and inadequacy of transport facilities. Thus, if the surplus crop could be conveniently and economically channeled into production of garlic powder by this improved technique, a considerable national waste could be avoided. Incidentally, this will also assist in regularizing the market rates, in particular, during the glut season
2. All undersized or culled, but, healthy bulbs, which fetch very low price in the market, could be conveniently utilized in the manufacture of garlic powder with advantage
3. Given the necessary publicity, garlic powder should prove very handy to the home makers for day -to-day use both as a condiment and as a folk medicine. It could also save the botheration of peeling garlic everyday, saving time

4. A cheaper garlic powder of good overall quality could be most welcome to the food industry in India for flavoring several food products
5. Garlic powder converted into suitable tablets or packed into capsules should find a steady market in India, in the face of expensive synthetic tablets claiming anti-microbial properties
6. Odorless garlic powder: Pruthi et al. (1959) also succeeded in the manufacture of an odorless garlic powder by inactivating the enzyme alliinase, naturally present in garlic. It had almost the same composition as that of normal garlic powder, except in respect of allicin, the pungent principle in garlic, which could be regenerated at will, by either incorporation of fresh enzyme or freshly crushed garlic in a small quantity (Pruthi 1980)
7. Garlic Salt: It is prepared by rapidly mixing in low humidity room the following:

   Garlic powder: 20 parts
   Refined pulverized salt: 78 parts
   An anti-caking agent 2 parts(usually calcium stearate)
   The composition will vary according to the garlic powder used. Both garlic powder and common salt are hygroscopic, hence, it is important for immediate packing in moisture-free containers like glass, plastic bottles, capped airtight. The storage defect of garlic powder is obviated by the development of a stabilized garlic powder with encapsulated garlic oil (Ramalakshmi et al. 1997).
   Oil of Garlic: It is prepared by steam distillation of crushed fresh garlic cloves. It's yield is very low (around 1%), and it is used in medicine/garlic capsules and other pharmaceutical preparations, apart from its use in flavoring various foods, meat preparations, soups, sauces etc. Its scope is at present limited. The flavor strength of garlic oil is 200 and 900 times that of dehydrated and fresh garlic respectively (Ferrel 1985). Garlic oil is reported to be an excellent substitute for fresh garlic (Heath 1978). For easy blending and mixing in seasoning mixtures, oil is available in different dilution levels at 5, 10 and 20% in vegetable oil (Ferrel 1985) and is also for preparing stabilized garlic powder or garlic flavor (Ramalakshmi et al. 1997).
   Garlic Oleoresin: Garlic concentrate is manufactured by extraction of garlic with water by vacuum concentration. Garlic oleoresin is described as a "dark brown soft extract". To confer homogeneity and ease of handling, the concentrate is usually compounded. Essentially garlic extract contains sugar, water, minerals, nitrogenous compounds and certain amount of aromatic materials. The essential oil constitutes the aroma, commercial oleoresin containing up to 5% essential oil (Ferrel 1985; Heath 1978). One part of garlic oleoresin is equivalent to 50% of fresh garlic. There are several types of garlic oleoresin extractives depending on end use (Table 3).

The containers of dehydrated garlic should be stored in covered premises, protected from the sun, rain and excessive heat. The storage room should be dry, free from unpleasant odors, and proofed against the entry of insects and vermin.

Table 3 International physico-chemical quality specifications/requirements for dehydrated garlic

Physico-chemical characteristics	Quality requirements	
	Garlic slices,rings, flakes, pieces	Garlic powder, gritts
Moisture content, % (m/m), max.	8	6
Total ash, %(m/m) (dry basis), max.	5.5	5.5
Acid-insoluble ash% (m/m) (dry basis) max.	0.5	0.5
Extraneous matter %max.	0.5	0.5

Source: ISO5560, 1983. International Standard for dehydrated garlic and Garlic Powder

Transport The containers of dehydrated garlic and garlic powder should be clearly marked with warning against careless handling which might lead to perforation of the containers. They should be kept dry and cool, stored well, away from ship's boilers or bilges, when transported by sea over long distances.

Uses of Garlic

Garlic has since long been recognized, all over the world, as a valuable condiment for foods, and also as a popular remedy for various human ailments and physiological disorders.

Garlic as a Condiment Garlic is practically used all over the world to impart flavor to the food preparations. In the USA, about 50% of the entire output of fresh garlic is dehydrated and sold to food processors for use in mayonnaise products, salad dressings, tomato products, and in several meat preparations. Raw garlic can be used in the manufacture of garlic powder, garlic salt, garlic paste, garlic cheese croutins, garlicked potato chips, garlic bread, garlic vinegar, garlicked meat tit-bits, and garlicked bacon etc.

Both in Europe and Latin American countries, garlic is very widely used as a condiment in various food preparations.

In India, and other South Asian countries, including the Middle Eastern countries, its wide use in the food industry is well recognized.

Garlic is also used, ground with other spices, sliced and fried or bruised and steeped in water, in vegetables, meat and fish preparations to act as an anti septic and to improve flavor and digestion. These properties make it a desirable ingredient for pickles.

Garlic has a strong characteristic odor and acrid taste. In most Mediterranean and Asian countries, garlic forms a part of almost all dishes, mainly to flavor meats, vegetables, stews, soups, salads, tomato dishes, sauces, and preparation of sausages and hot pickles.

Oil of garlic has been appreciated as a valuable flavoring agent for use in all kinds of meat preparations, soups, canned foods and sauces on a commercial scale.

Garlic as an Insecticide

Recent studies at the famous Bhabha Atomic Research Center, Mumbai, India, has shown that there is differential toxicity of garlic vapor to housefly and "*khapra*" beetle. Hyper excitability, atexia, salivation and excretion – the usual signs of poisoning when exposed to chemical insecticides – has been shown with garlic vapor.

Garlic Paste/Mixture as Biofungicide A garlic remedy for the dreaded "slow wilt" disease in Black pepper is described by Shanthimoole and Pailoor (1991).

Garlic Oil as an Adhesive Garlic oil is highly volatile. It is very effective as an adhesive for broken glass.

Garlic Residue with Antibacterial Properties According to an American patent, the residue of garlic, obtained by alcoholic extraction and distillation, contains bacteriostatic and bactericidal substance, identified as "allyl disulfide oxide".

On steam distillation of crushed garlic at atmospheric pressure, the major color producing principle, "allicin" decomposes to diallyl disulfide and other disulfides. That explains why the volatile oil of garlic consists of chiefly disulfides. However, "allicin" can be isolated by steam distillation at a reduced pressure. It may also be clarified that the mother precursor allicin does not possess bactericidal properties. It is only allicin produced as a result of enzyme cleavage of alliin which is bactericidal.

Garlic in Medicine

The medicinal qualities of garlic have been celebrated since thousands of years. Pliny says that it is a remedy for 61 ailments, while Aristotle, Sotion and Dioscorides have sung its praises.

Garlic has been used to cure dropsy, pherenitis, jaundice, neck swellings, bronchitis etc. Externally garlic is used as a remedy for skin and ear diseases. In western medicine also, it is used to a fairly large extent. Garlic juice mixed with 3 or 4 parts of ordinary or distilled water has been used as a lotion for washing wounds.

The inhalation of garlic oil or drinking garlic juice has been found to be useful in combating pulmonary tuberculosis, rheumatism, sterility and impotency. Garlic juice is used for various ailments of the stomach. Also, as a rubefacient for skin diseases and as ear drops for ear ache. The juice diluted with water can be used against duodenal ulcer. In Cambodia, garlic leaves are used to treat asthma.

On account of its highly curative properties, garlic is described, in ancient Indian texts, as derived from *Amrita* or Ambrosia. The prejudice against garlic among some Indians, especially of the *Brahmanic* class, is because of its association with alien invaders of India. Recent scientific investigations clearly point to the remarkable properties of garlic and it is, indeed, a cheap folk medicine in the hands of the house wife.

Clinical application of garlic in 114 human patients with clear hypertension and atherosclerosis showed marked improvement in their condition. It was interesting to note that after treatment, the systolic arterial tension dropped by 8–33 mm and diastolic arterial tension by 4–20 mm.

In small doses, garlic intensified the intestinal muscle tone, thus enhancing peristaltic movements, while in larger doses it inhibited them. In experiments on animals and human workers, exposed to chronic lead intoxication, garlic has been found to be an efficacious preventive ad curative drug against saturnism. In the ancient Indian Ayurvedic system of medicine, garlic has been found to have very positive and beneficial effects in over 18 human diseases.

Garlic as a Cure for Cancer Garlic antibiotic allicin has been found effective against malignant tumors of human as well as animal cell. Transplanted tumors and Jensen sarcoma in rats regressed, and in some cases, completely disappeared after injection of 1–3 mg of allicin, given either intramuscularly or directly into the tumor. However, all types of tumors do not respond equally effectively. This aspect warrants further scientific investigation.

Garlic in Human Nutrition There is an increase in Vitamin B 1 synthesis, on the administration of garlic to human subjects; as alli-thiamine is formed by the reaction of thiamine (Vit B1) and allicin, which is also absorbed faster than thiamine in the intestine. Thus, there appears to be a fundamental scientific basis for the therapeutic use of garlic and also on the utilization of food ingredients. It also provides a scientific explanation for its popularity as a folk medicine for over thousands of years, in India, China and Egypt.

Dixit and Joshi while substantiating the earlier claim that garlic decreases blood cholesterol, and blood sugar, have demonstrated that beneficial actions are associated with testicular damage in stoppage of sperm genesis, if ingested for long periods over 75 days.

References

Anwar MM (1988) Packaging of garlic and ginger: Souv. Symp Packaging of Fruits and vegetables Agri Hort Soc Hyderabad:1–5

Bottcher H, Gunther I (1994) Quality changes of dry garlic during the log-term storage. 1. External Quaity Nahrung Food 38(1):61–70

Cosenza GW, Menezes SJADE, Regina SM, Gontizo UPM (1981) Pesquisa Agropecuria Brasileira 16:199–203

Directorate of Marketing and Inspection, Government of India, Faridabad (1968) Marketing of Minor Spices (including garlic) in India, Marketing Series No. AMA, 163. Manager of Publications, Delhi, pp 1–228

Ferrel KT (1985) Spies and culinary shrubs. Avi. Pub, Westport, pp 114–120

Garcia Lopez C (1993) The mechanization of the cultivation and harvesting of garlic. Maquinas Y Tractores Agriccolas 4(6):56–60

Heath HB (1978) Flavour technology. Avi. Pub. Co., Westport, pp 175–178

Nair KPP (2019) Intelligent soil Management for Sustainable Agriculture – the nutrient buffer power concept. Springer, Cham

Peter GB, Selvaraj N, Vedamuthu, Pillayarsamy (1995) Rubbering and premature sprouting of garlic bulbs. Spice India 8(7):11–12

Priyadarshan PM, Ramakrishnan VV, Madhusoodanan KJ (1989) Spice India 2(4):6–7

Pruthi JS (1980) Spices and condiments – chemistry, microbiology and technology, 1st edn. Acad. Press, Inc, New York, pp 1–450

Pruthi JS (ed) (1987) Bibliography of 57 spices and condiments. (1972–1986). AFST, Delhi, pp 1–168. (1617 references, including 88 references on Garlic)

Pruthi JS (1994) Innovations in post-harvest technology of spices. In: Chadha KL, Rethinam P (eds) Advances in horticulture Vol 10, Plantation and spice crops, part 2, 1st edn. Malhotra Publishing House (MPH), New Delhi, pp 1255–1282

Pruthi JS, Singh LJ et al (1959) Special "garlic powder" number- covering a series of 10 research papers on different aspects of quality, technological aspects of manufacture, packaging requirements and storage of garlic powder. J Food Sci 8(12):429–464

Ramalakshmi R et al (1997) Effect of storage on the quality of "stabilized" garlic powder. Indian Spices 34(4):5–7. 19

Setty BSK, Sulikeri GS (1990) Effect of NPK fertilization on loss of weight on garlic bulbs during storage. Spice India 3(1):9–10

Shanthimoole MT, Pailoor S (1991) Spice remedy for spice disease. Spice India 4(2):24

Singh LJ, Pruthi JS, Sreenivasamurthy V, Lal G (1959) Effect of regional variability in garlic on the quality of garlic powder. J Food Sci 8:431

Leek/Stone Leek/Welsh Onion

Botanical names: (1) *Allium porrum* Linn., Leek or Winter leek (2w)*Allium fistulosum* Linn. Sone leek or Welsh onion
Family: Liliaceae

Description and Distribution Leeek is a non-bulb forming member of the onion family and is grown for its blanched stems and leaves. Distinction must be made between "Stone leek" and "Winter Leek", both being members of the onion and garlic family, Liliaceae. Stone leek is *Allium fistulosum,* Welsh onion or Ceboule or Japanese bunching onion, while winter leek is *A. purrum*. It is reported that sometimes Welsh onion is used as a substitute for leek. *A.fistulosum* has been mentioned in Japanese literature as early as 918 AD. In Europe, USA and India it is grown mostly in home gardens.

"Stone leek" or Welsh onion is a hardy perennial, but, can be grown as an annual or biennial. It does not form a real bulb, but, there is only a small enlargement at the base. Propagation is done by division or seeds, latter mode is preferred.

Economic Importance of Dehydrated Leek Dehydrated leek constitute the fifth most important bulb crop imported into Europe.

Soil and Climate Both soil and climatic requirements are the same as of onion and garlic. Leeks are best grown in areas with cool temperature.

Common Varieties American Flag and London Flag are the most common varieties grown. No such information is available about in India.

K. P. Nair, *Minor Spices and Condiments*,
https://doi.org/10.1007/978-3-030-82246-0_4

Crop Management

Raising Nursery and Transplanting Seeds are sown earlier than onion sometimes at the end of September at 5 kg/ha. Transplanting method is same as for onion except wider spacing at 15 cm. Transplanting is done in trenches about 30–45 cm deep.

Manuring As leek plants are larger than that of onion, higher fertilizer requirement is needed. A 30 tonne crop removes 100 kg N, 65 kg P and 130 kg K per hectare. Other operations are same as for onion.

Seed Production It is a biennial crop and the seeds are produced in India at higher altitudes on hills with cooler climate.

Harvest At harvest, Leeks should have green fresh tops and medium-sized necks. The entire plant is pulled out of soil and the leeks are packed in a crate in a manner similar to that used for celery.

Plant Protection Same as in the case of onion (Table 1).

Leek Essential Oil New 85 volatile compounds have recently been isolated from leek essential oil in Germany (Stephani and Baltes 1992).

Post-Harvest Technology and Processing Leeks are perishable and should be processed as soon as possible after harvest. If storage is necessary, a temperature of 32 degrees centigrade and a RH of 90–95% is ideal.

Table 1 Composition of Leek (*A.porrum*)

Property	Content
Moisture	78.9%
Protein	1.8%
Fat	0.1%
Minerals	0.7%
Carbohydrates	17.2%
Calcium	0.05%
Phosphorus	0.07%
Iron	2.3 mg
Vit A	30 I.U
Vit B1	0.23 mg
Vit C	11 mg/100 g
Calorific value	77 calories/100 g
Combined sulfur	0.06–0.072%

The leeks are unloaded onto a continuous conveyor belt, and so oriented, that as they pass by a revolving blade, the roots are sliced off. A thorough washing is important to remove soil particles, dirt and silt that collects in the crotch of the leaves. The bulbs and leaves are diced together into 3 inch dice. These are loaded onto trays at 18 lb. or less per square foot. A sulfite/bisulfite mixture is sprayed over the dice in a concentration to give 500–1000 ppm of SO2 (2 substrate please) in the dehydrated product. Drying temperatures are similar to those previously discussed for onions.

Uses Like the green bulb onion, stone leeks are used for flavoring or for livening dull starchy diets of eastern people of India. They are also used in salads or eaten raw alone or used as a flavoring agent in soups and stews. They are mild flavored. The tender leafed varieties are raised for their edible tops while others, for their white blanched leaf bases.

Only scarce information is available in India on their technological utilization, volatile oil, sulfur compounds and their antiseptic properties, as is so well known about onion and garlic, the elder brothers in the family. However, some useful information is available elsewhere in western countries.

Reference

Stephani A, Baltes W (1992) New aroma compounds of leek oil. Zeitschrift fur Lebensmittel Untersuchung und Forschung 194(1):21–25

Ajowan or Bishop's Weed

Botanical name: *Trachyspermum ammi* (Linn) Sprague Syn. *Carum opticum* Heim

Description and Distribution *Ajowan* or *ajowain* is an annual herbaceous plant, bearing greyish-brown fruits (seeds), which constitute the spice. It is an erect, glabrous or minutely pubescent, branched annual, up to 90 cm tall, cultivated almost throughout India. Stems are striate, leaves are rather distant, 2–3 pinnately divided segments linear and ultimate segments are 1.0–2.5 cm long; flowers are in terminal or seemingly terminal lateral pedunculate, compound umbels, white; fruits are small, ovoid, muricate, aromatic cremocarps, 2–3 mm long with greyish brown compressed mericarps with distinct ridges and tubular surface,1-seeded. It belongs to the family Umbelliferae, comprising 270 genera and 2700 species, mostly grown in the temperate regions of the world, excepting for a few species, which are cultivated in the tropics. S.W. Asia (Iraq, Iran, Afghanistan and Pakistan) and especially North India are the regions where it is widely grown. In India, it is largely grown in the Uttar Pradesh, Bihar, Madhya Pradesh, Andhra Pradesh, Punjab, West Bengal, Rajasthan, and Tamil Nadu. It is generally grown in October–November and harvested in May–June.

Area, Production and Export Large quantities of *Ajowan* are grown in India and the entire demand is met by indigenous production. Several commercial types of ajowan are known, of which *desi ajowan* (large), *desi ajowan* (small) and *Nadiad ajowan* are the most common. Some of the best quality, small-seeded varieties are produced in India, in Ujjain and Gwalior in Madhya Pradesh. As the crop is grown to a large extent in small fields, authentic data regarding the area and production is not available. However, considerable quantities of *ajowan* are exported from India.

Varieties and Selections At the Indian Agricultural Research Institute, (IARI), New Delhi, India, Joshi et al. (1963) compiled the field trial results of *ajowan* at four locations (1) Katrain, Kulu Valley (2) IARI farm (3) Pusa (Bihar State) and (4)

K. P. Nair, *Minor Spices and Condiments*,
https://doi.org/10.1007/978-3-030-82246-0_5

Jamnagar (Gujarat State) and reported that Selection Numbers S 25, S 95, S 47, S 84 and IC 3743 yield essential oil up to 3.9% in plains and up to 4.9% in cooler areas (Katrain). They also observed considerable variations in agronomic characteristics of the varieties as follows:

(1) Number days to flower	82–132
(2) Number of seeds per umbel	150–345
(3) Weight of 100 seeds	0.1–0.32 g
(4) Single plant yield	3–23 g
(5) Essential oil content	1.28–4.08%

Srirama Rao (1982) and Srirama Rao and Marunakara Basu (1979) recommended the following selections, out of 46 screened, namely, (1) Lam-Sel- I (erect, early maturing 135 days) with a yield potential of 8 quintals/ha and (2) Lam-Sel- II (bushy, late maturing 160 days) with a yield potential of about 10quintals/ha.

Depending upon the rainfall, the crop can be sown from mid August to mid October, using a seed rate of 3–4 kg/ha.

Crop Management

Climate and Soil The crop is grown during cold weather, both as a dry crop and under irrigation. In heavy soils, it is also grown as a rainfed crop, grows on all types of soils, but, does well on loams or clayey loams. For the dry crop, the black cotton soils, which store moisture of the earlier rains are very suitable. Under irrigation, it is grown extensively as a garden crop or in small fields. Although it is largely grown in the plains, it also comes up well in the dry Deccan plains of India. Heavy black cotton soils are well suited for ajowan (Gopinath and Poonacha 1992).

Cultural Practices

Irrigated Crop Cultivation Sowing is between May and November in different parts of the country. Seeds are broadcast in the soil at the rate of 2.5–3.5 kg/ha. Germination takes 5–15 days, light irrigation is given immediately after sowing and thereafter regular light irrigation is given and weeding, at least two. At the first weeding the crop is thinned, optimum spacing is 45 cm × 30 cm. Flowering takes place in about 2 months and at this time field looks like an ornamental garden, with delicate snow – white flowers, borne in large umbelliferous clusters. As the plant branches profusely, the umbels mature unevenly. There is considerable interval between the ripening of the first-formed umbels and the later ones on the secondary or tertiary branches.

Harvesting: Extends from February to March Fruits are ready for harvest when flower heads turn brown. Plants are pulled out by roots, dried on mats on the threshing floor. When completely dried, fruits are separated by careful rubbing by hands or feet followed by winnowing. Yield, on average, varies from 3 to 3.5 quintals/ha, though higher yields of 8–10 quintals are also reported.

Rainfed Cultivation

As a rainfed crop, ajowan is grown extensively in the black cotton soils of Andhra Pradesh, Maharashtra, Gujarat, Andhra Pradesh and Karnataka. Generally, the seed rate is much higher than that for irrigated crop. A dry crop yields just about 2–2.5 quintals/ha. Sowing is done in October–November and harvest in end March.

Cultivation as a Garden Crop Grown as a garden crop in small fields. On a commercial scale it is grown mostly mixed with other crops, such as, Indian mustard (*Brassica juncea*) or pearl millet or sorghum as in Andhra Pradesh.

Pest and Diseases Ajowan is not, generally, affected by many pests and diseases. However, collar-rot and root-rot caused by *Sclerotium rolfsii* Sacc. have been recorded. Ajowan treated with a number of common fungicides and insecticides and stored in paper bags at room temperature (26 degrees to 37 degrees celsius) and RH 83% resulted in higher germination percentage of the fruit. In ware houses, cold treatment at 2 degree centigrade for 16 days destroys many fungus (Thomas 1965).

Post-Harvest Technology (Table 1)

Indian standards have graded ajowan into the three following classes:

1. Special
2. Good
3. Fair

Adulterants Commercial samples of ajowan are often found to be adulterated with earthy matter, seeds of *Makara*. In an analysis of the market sample, the total ash content varied from 7.2% to 9.4% and acid insoluble ash from 0.45% to 1.5%.

Packaging and Storage Ajowan seeds free from leaf, chaff, stones etc., are packed in clean gunny bags (40–45 kg capacity) and stored in hygienic warehouses/go downs; normally, they are stored for 6–8 months, but, can be kept up to an year.

Table 1 Composition of Ajowan

Property	Content
Moisture	8.9%
Protein	15.4%
Fat(ether extract)	18.1%
Crude fiber	11.9%
Carbohydrates	38.6%
Mineral matter (total ash)	7.1%
Calcium	1.42%
Phosphorus	0.30%
Iron	14.6 mg/100 g
Calorific value	379/100 g

Processed Products

The major processed products are the following:

1. Oil of ajowan, prepared by steam distillation
2. Thymol, produced by treatment of the ajowan oil with aqueous alkaline solution and regenerating thymol from it by extraction with ether or steam distillation.
   Both the above products are used in medicine/pharmaceutical industries. For thymol there is Indian Standard (ISI).
3. Dethymolized oil or thymine for industrial purposes
4. Fatty oil, produced by method developed by the Oil technological Research Institute, Anantapur (Andhra Pradesh).

Essential Oil

Umbelliferous seeds, such as, ajowan, coriander, cumin, dill and fennel contain 2 kinds of oils namely, (a) essential oil or volatile oil and (b) non-volatile fatty oil. Two integrated methods have recently been developed by Ramachandraiah et al. (1988) to recover both oils from crushed seeds.

Uses Oil of ajowan is an almost colorless to brownish liquid possessing a characteristic odor and sharp burning taste. On standing, a part of the thymol may separate from the oil in the form of crystals which are sold in the Indian market under the brand name of *ajowan ka phul* or *sat ajowan*, which is highly valued in medicine as it has all the properties ascribed to ajowan seed. This was used in surgery as an antiseptic and was also found to be of great value in the treatment of human hook worm disease, and appeared to be far more effective than wormseed oil (*Chenopodium ambrosioides).* This product was shown by R. Haines in 1856 and J. Steinhouse in

1855 simultaneously, to be identical to thymol which was previously obtained from thyme (*Thymus vulgaris*), and was one of the solid essential oil fractions obtained in the seventeenth century. The aqueous solution of thymol is an excellent mouth wash and thymol is a constituent of many toothpastes. The distillation water, the essential oil, and the thymol-separated therefrom are used in India in medicine, in particular, to treat cholera. In India, the ajowan seeds are used in making the greatly popular "*paan beeda*" (a combination of betel leaf, arecanut and lime to obtain red color, chewed frequently) of northern India.

Medical Use

Ajowan seeds are useful in treating flatulence, colic, atonic dyspepsia, diarrhoea, cholera, hysteria and spasmodic afflictions of bowels. The seeds produce a feeling of warmth and relieve feelings of sinking and fainting, which accompany bowel disorders. They are most frequently used in conjunction with asafoetida, myrobalan and rock salt. Externally, ajowan is applied in mixtures to relieve rheumatic and neurologic pain. A teaspoonful of ajowan seeds with a little salt is a common domestic remedy for indigestion from irregular diet. For stomach ache, cough and indigestion, the seeds are masticated, swallowed and a glass of hot water is drunk. They are also useful in treating skin diseases. A hot poultic of seeds is used as a dry fomentation to the chest in treating asthma and applied to hands and feet in controlling cholera and fainting. They check chronic discharge from bronchitis. Extracts of seeds in 70% and 40% alcohol are toxic to *Staphylococus* and *Escherichia coli*, the former in most infections and the latter in urinary tract infection (UTI), in both males and females. The leaves are used as a vermicide. Ajowan is one of the most effective spices in medical therapy, thereby significantly enhancing human health in a number of ways. The roots of ajowan plant have both diuretic and carminative properties.

Comparative Antimicrobial Activity of Seed Spice Essential Oils: Antimicrobial activity of spice essential oils against a range of microorganisms was investigated using paper disc diffusion method. Spice essential oil of cumin, ajowan, ginger, coriander, basil and clove and also eugenol and allyl-isothiocyanate (volatile oil of mustard) were used. *Lactobacillus acidophilus, Bacillus cereus, Saccharomyces cerevisiae, Mycoderma* sp. and *Aspergillus niger* were used as test organisms. The essential oils showed various degree of inhibition against test organisms. *Mycoderma* sp. was the most susceptible and *Bacillus cereus* was the most resistant to all spice essential oils. Susceptibility followed the order: *Bacillus cereus> Lactobacillus acidophilus> Saccharomyces cerevisiae>Aspergillus niger> Mycoderma sp.* Greatest antimicrobial activity was observed for allyl-isothiocyanate followed by oil of ajowan and eugenol, both at ambient temperature and at 37 degree centigrade. The oils from other spices seeds were less effective (Meena and Sethi 1994).

References

Gopinath G, Poonacha NM (1992) Bishop's weed. Spice India 5(9):9–14

Joshi BS, Ramanujam S, Saxena MBL (1963) Improvement of some essential oil bearing spice plants. PRL Bull 1(2):94–100

Meena MR, Sethi V (1994) Anti-microbial activity of essential oils from spices. J Food Sci Technol 31(1):68–78

Ramachandraiah OS, Azeemoddin G, Thirumala Rao SD (1988) Integrated methods of obtaining essential and fatty oils from umbelliferous seeds. Indian Perfumer 32(1):55–60

Srirama Rao T (1982) Recent advances in minor spices improvement work in Andhra Pradesh. Seeds Farms 8(5):22–24

Srirama Rao T, Marunakara Basu M (1979) Improvement of minor spices in Andhra Pradesh. Indian Cocoa Arecanut Spices J 3(1):3–4

Thomas PM (1965) Warehousing of spices in India, Spices Bull. (Seminar no 4 (10),61). Spices Promotion Council, Kochi

Celery Seed

Botacal Name: *Apium graveolens* Linn. var.Celery
Family: Umbelliferae

Description and Distribution Celery (*Apium graveolens* Linn.) is an umbelliferous aromatic herbaceous plant grown for its leaves, seeds, oleoresin and essential oil. Celery seed has been known for over 3000 years. It was mentioned in Homer's Odyssey, written about in the seventh century BC, as an excellent medicament. The earliest recorded use of celery seed as a seasoning for food was not until 1623 in France.

The native habitat of celery extends from Sweden to Egypt, Algeria and Ethiopia and in Asia, India, Caucasus and Baluchistan. It is also grown in France and in China.

It is also claimed to be the first mentioned as a cultivated food plant in France in as early as 1623. It is cultivated both for use as salad and seed-raising in the northwest Himalaya, Punjab, Uttar Pradesh and Haryana and elsewhere in France and in USA. In Amristar district alone in Punjab state about 1200 ha has been under this crop in the *rabi* season as early as 1969–1970. In colder climate and in hills, celery grows as a biennial plant and produces seeds only in the second year, but, in plains, it becomes an annual and produces seed in the first year itself, about 4–5 months after transplanting. Indian celery seed plant is about 1.5 m (4–5 feet) feet tall, while that grown as a vegetable/leaf/stalk it is only less than 2 feet tall.

In India, its commercial importance is in seed production, which has a ready market. Of late, its cultivation has extended to the hilly areas of Jammu and Kashmir. Experiments conducted at the Regional Research Laboratory (RRL) in Jammu has shown that the crop can successfully grow at high altitude and cold climate.

K. P. Nair, *Minor Spices and Condiments*,
https://doi.org/10.1007/978-3-030-82246-0_6

Economic Importance and International Trade

India dominates the world export of celery seed. By far, the most important market for Indian celery is the USA which singly imports about 75% of the celery exported from India. Maximum export of celery from India to USA was in 1993 at 3108 tonnes which declined to 2704 tonnes in 1994. The top five countries to which India exports are as follows (Table 1).

Area and Production

Punjab State in India is the one where most celery is grown, and, the district of Amritsar leads the others, like Jalandhar, Ludhiana, Ladwa and Saharanpur.

Varieties The Indian Agricultural Research Institute, New Delhi, India, has recommended the following varieties of celery for cultivation for salad purposes:

1. Standard Bearer: Early maturing, medium tall plant, medium pink stems with white longitudinal streaks, stalks solid and of good size and flavor
2. Wright Grove Giant: Medium late maturing, tall growing plants, produces large white stalks of fine quality; immense cropper

There is but minimal published information on celery varieties recommended for seed production in India. There is a distinct difference in flavor of celery plants producing celery leaves, flakes have sweet, strong typical celery aroma and flavor, whereas seeds have a warm and slightly bitter flavor and taste.

Crop Management

Celery can be grown on a wide range of soils, but, thrives best on fertile loam soils. Heavy clayey soils are unsuitable. Celery can follow early potato (mid September-mid December) and rice (July–November) crop successfully. The vegetable growers

Table 1 Export of Indian celery – countrywise performance (latest figures)

Country	Value in US dollar (million)
1. USA	7.43
2. Netherlands	0.75
3. South Africa	0.75
4. Australia	0.29
5. Canada	0.29

transplant celery on the ridges in the standing crops of cauliflower and tomato to save it from frost. It requires moist soil and cool climate.

Land Preparation

As celery is an exacting crop, soil should be well prepared (pulverized) and seed bed should be heavily manured. Heavy application of farm yard manure is recommended. For easy irrigation, bunds around should be provided in the field.

Sowing Method

500 g seeds are used to raise nursery. One kg each of calcium ammonium nitrate and superphosphate are added to the nursery beds at the time of sowing seeds. Irrigation water is applied through a sprinkling can after every 4–5 days until seeds germinate in about a fortnight. The nursery beds may be irrigated once a week later. Nursery seedlings would be ready for transplanting in the second week of December. Irrigation water must be stopped to the nursery a week before transplantation. But, nursery beds should be irrigated 12 hours before transplantation. In the main field transplantation should be done at 45 cm between rows and 30 cm between plants. This operation should preferably be done in the evening and an irrigation should follow the next day.

Fertilizers

Calcium ammonium nitrate 200 kg/ha and superphosphate 100 kg/ha should be drilled into the soil before transplanting. If the soil test shows K deficiency 30-40 kg/ha of Muriate of Potash is also applied. In soils normally fertile, K residue left from previous crops of potato or rice will suffice. The second dose of 200 kg calcium ammonium nitrate may be applied in March, just prior to or during emergence of the flowers. In the case of silage farms, only superphosphate need be applied.

Irrigation

One irrigation every fortnight should be given in December, January, February and March. When the weather warms up in April and May, weekly irrigation may be given. The crop is liable to lodge if fierce winds blow after irrigation. Two or three hoeing are done to get good growth of plants. First hoeing is done about 3 weeks after transplantation and subsequent hoeing 2 weeks after each hoeing. The crop responds well to hoeing as it invigorates root growth.

Plant Protection

Punjab has not reported any disease attack so far. The crop is, however, susceptible to certain fungal diseases such as early blight and late blight. Spray of Bordeaux mixture controls the former and sulfur dust the latter. No insect attack is observed, possibly due to the repulsive odor from the alkaloids the plants emit.

Tissue Culture

The major breakthrough in tissue culture in celery is the encapsulation of somatic embryos from hypocotyl segments, which leads to the production of artificial celery seeds. Priyadarshan et al. (1989) have reviewed the subject.

Threshing and Harvesting

In early May the flowers turn reddish. This is the right time for harvest. The crop should not be left in the field for long, when the seeds will overripen. Harvesting at the right time will prevent seed shedding. Plants which are very light are likely to be blown away by the wind. Harvested crop is threshed using sticks the very next day and seeds stored in gunny bags in the go downs. Winnowing and sieving is done by contract laborers from Jammu. Seed yield is about 5–6 quintals/acre.

Post Harvest Technology

Cleaning/Preparation for the Market:

Harvested seeds after threshing usually contains 10–15% extraneous matter (dust, small pebbles etc.) and organic matter (stalks, straw, leaves etc.) as against the permitted PFA limit of 1–2%. They are sieved out, winnowed and cleaned further before marketing (Table 2).

Table 2 Composition of celery seed and celery leaf and stalk for salad

Property	Content	
	Seed	Leaf and Stalk
Moisture	5.1%	81.3%, 93.5%
Protein	18.1%	6.0%, 0.8%
Fat (ether extract)	22.8%	0.6, 0.1%
Crude fiber	2.9%	1.4%, 1.2%
Carbohydrates	40.9%	8.6%, 3.5%
Total ash	10.2%	
Calcium	1.8%	0.23%, 0.3%
Phosphorus	0.55%	0.14%
Iron	0.45%	0.06%, 0.05%
Sodium	0.17%	
Potassium	1.4%	
Vit B1 (thiamine)	0.41 mg/100 g	Trace
Vit B2 (riboflavin)	0.49 mg/100 g	
Niacin	4.4 mg/100 g	
VitC (ascorbic acid)	1.7 mg/100 g	62.6 mg/100 g
Vit A	650 international units(IU) 5800, 7500 IU	
Calorific value (food energy) 450 calories/100 g	64.18 cal/100 g	
Total solids	18.7%,6.5%	
Mineral matter	2.1%,0.9%	

Adulteration in Celery Seed

Celery seed is marketed whole and in ground form. Adulteration is done by addition of exhausted or spent seeds (from which oleoresin or oil has been extracted), excess stems, chaff, soil and dust. The ground celery is sometimes adulterated with farinaceous products, linseed meal, worthless vegetable seeds or, at times, even with weed seeds.

Grading

The ISI has laid down Indian Standards for celery seeds into three grades as follows:

1. Special
2. Good
3. Fair

Packaging and Storage

Well dried ad cleaned celery seeds are packed in 50 kg hessian bags and transported to terminal markets for sale or to warehouses for storage.

The American Spice Trade Association (ASTA) has prescribed the tolerance limits of contaminants as follows for celery seeds (Table 3).

Processed Products

Seed-Based Processed Products

Types of Essential Oils from Celery

Two essential oils, celery herb oil and celery seed oil are distilled from celery seeds. The former is the true herbaceous oil, but, its commercial prospects at the present is rather low. The latter is produced from celery seeds or fruits alone and briefly discussed below.

Celery Seed Oil Aroma of celery seed oil is spicy, warm, slightly fatty, fruity, penetrating and very persistent. Its flavor is similar in character, but, has a burning sensation and is very bitter.

Celery seeds on steam distillation yield 2–3% of a pale-yellow volatile oil with a persistent odor, characteristic of the plant. In trade, this is known as "celery seed oïl" and is much valued both as a fixative and as an ingredient of novel perfumes, pharmaceutical preparations and medicines (Tables 4 and 5).

Celery herb (Leaf) Oil The volatile oil content of celery leaves is too low to permit their economic distillation and hence it has not attained any commercial importance as of now. However, the celery herb oil is quite different from celery seed oil, the former being more representative of the plant, as such. Undoubtedly, this new type of herb oil could be introduced to great advantage in many products of food industry, notably soups to impart fresh celery leaf flavor to advantage.

Table 3 Tolerance limits for contaminants

Contaminant	Tolerance limit
Whole dead insects by count	4
Mammalian excreta(mg/lb. seed)	3
Other excreta (mg/lb)	3
Mold (%by weight)	1
Insect defiled/infested (% by weight)	1
Extraneous matter (% by weight)	0.50

Table 4 Celery seed oil constants

Specific gravity	0.850–0.875
Refractive index	1.478–1.486
Optical rotation	+65 degrees - 82 degrees
Acid value	Up to 5
Ester value	15–40

Table 5 The principal constituents of celery seed oil

Property	Content
d-limomonene	60%
d-selinene	10%
Sedanonic acid anhydride[a]	0.5%
Sedanolide[a]	2.5–3.0%

Note: [a]Responsible for the aroma of the oil

Celery Chaff Oil It has a somewhat harsher and coarse odor and flavor than that of celery seed oil. Celery seed oil is frequently adulterated with celery chaff oil or with d-limonene, the addition of which is difficult to detect. It has been reported that the odor and flavor on Indian celery seed oil is normal, but, not usually quite as fine as those of French celery seed oil, which fetches a much higher price in the global market.

Celery Seed Oleoresin

The celery seed oleoresin is a green liquid having a volatile oil content of 9 ml/100 g. The oleoresin of celery seed is of two types (1) French (2) Indian. The French type is sweet, herbal and tenaceous with only a slight citrus undertone, whereas the Indian type is more herbal with a slight lemon-like aroma and tenacious herbal undertones. The flavors are slightly different, with the French product being more pleasing; however, both are quite bitter.

Oleoresin of celery seed is prepared by extraction of crushed, dried celery seed with suitable solvents like food grade hexane or ethylene dichloride, filtration and desolventization under vacuum. The oleoresin not only possesses the volatile top note of the celery essential oil, but, also the "body", that is, the fixed or non-volatile extractive matter of the celery seed. Oleoresin could rightly be considered as "liquid celery seed" which is easier to handle in the preparation of tinctures and extracts.

Advantages of Celery Seed Oleoresin:

1. More convenient to use and store
2. Help maintain cleanliness
3. Easy to store
4. Imparts uniform standard flavor to the cold product
5. Microbial and Fungal contamination is eliminated
6. No enzymatic problem
7. No fumigation is necessary as is the case with storage of spices and seeds
8. Color distribution is also of uniform shade or proper mixing
9. More stable than whole or ground spices.

Celery Salt Commercial celery salt is prepared by mixing finely ground table salt with ground celery seed or celery seed oleoresin or ground dried celery stems. According to Canadian standards, celery salt should be a combination of 25% of celery seed powder or ground celery and Table salt (75 t%).

Celery pepper It substitutes ground black pepper for celery salt, but should not contain more than 70% pepper.

Processed Products from Celery Leaves and Stalks

Several products from dry ad dehydrated leaves/stalks have been in the market in the USA and other countries They are:

Dehydrated Celery It is commercially available as (1) 10 mm celery stalk dice (2) leaf and stalk flakes (3)stalk and leaf granules and (4) celery powder. These various styles of dehydrated celery products are used to flavor soups, for broth base, canned tuna fish, stuffings and stewed tomatoes and as a garnish on potato salad and meat sauces. Medium to dark-green celery is manufactured on a commercial scale in the USA.

Celery Stalk Products

Only stalk products have the deepest green color, and thus are frequently protected by the addition of minute amounts of sodium bisulfite and sodium sulfite. Celery flakes are used in dry soup mixes, canned soups, sauces, stuffings, casserol products and vegetable specialities. Granulated or powdered celery is a good choice for canned and frozen sauces and dry mixes for breadings and soups.

Processed Celery Juice Blends Celery-tomato juice blends from fresh celery leaves and tomatoes, successfully marketed.

Freeze dried celery made from cross-cut slices of celery stalks are also marketed in freeze-dried form. This process is particularly effective in retaining celery's original shape and crispiness. This makes a crisp garnish for potato salad, casserols, Chinese dishes etc.

How to Handle and Store Celery Seed Ad Its by Products?

Ground celery seed and extractives (oil and oleoresin) as well as dehydrated and freeze-dried celery should be stored in a cool, dark and dry place in order to protect the original color, flavor and aroma. Record dates on each container as they arrive in the store so that older stocks could be disposed off first. Store them off the floor and away from the walls to minimize changes in the dampness of the room. All containers must be tightly closed after each use, since prolonged exposure to air/atmosphere causes loss of flavor, color and aroma.

Uses

Celery seed and its byproducts find the following uses besides seeds serving as "bird feed".

As a Food Flavorant
The dried ripe fruits (seeds) are used as a spice. Leaves and stalks are used as salads and in soups as a pre-dinner appetizer. Celery is used as a natural medicine in different cultures throughout the human history. In Egypt wild celery seed is used as a diuretic, digestive aid, emmenagogue. It is also recommended as an aphrodisiac and anti-lactogen for nursing mothers (an anti-lactogen is any substance that suppresses lactation or the secretion of milk). In the indigenous system of medicine, for example Ayurveda in India, the oil is used as a nerve tonic and in digestive mixtures.

In Medicine
Seeds are stimulant and toxic and used in asthma and for liver diseases. Essential oil contracts graved and virginal uterus. It is also a remedy for rheumatism. Celery seeds are the main ingredient for celery tonics.

In Perfumery
The commercial use of celery seed oil is as a fixative; as an important ingredient in novel perfumes. Celery seed oil is one of the most valuable flavoring agents as it imparts a warm, aromatic and pleasing flavor to food products.

Fatty Oil
The fruits also yield about 7% of a fatty oil – "oil of celery". This is used as an antispasmodic and nerve stimulant. It has been successfully employed in rheumatoid arthritis and, probably, acts as an intestinal anti septic.

Celery Root

It is a diuretic and is administered in anasarca and colic. Celery seed serves also as a "bird feed".

Reference

Priyadarshan PM, Radhakrishnan VV, Madhusoodanan KJ (1989) Tissue culture of spice crops. Retrospects and prospects. Spice India 2(4):15–16

Coriander

Botanical name: *Coriandrum sativum* Linn.
Family: Umbelliferae

Description and Distribution

Coriander seed (botanically a "fruit", comprising two locules containing one seed each and fresh coriander (*dhania*, Indian Hindi name) leaves are too well known to need any introduction or description in India, in general, to the house wife/home maker, in particular, as Coriander seed is used almost daily in scores of Indian curries, to flavor them and its leaves for garnishing. Coriander is actually the housewife's secret of tasty food dishes.

Origin and Brief History

The origin of coriander (green leaf and seed/fruit) is virtually lost in antiquity. However, coriander has been used by man since long as coriander seeds were found in the Egyptian mummies/tombs (3000–4000 BC). Further, Persia grew coriander 3000 years ago and it added fragrance to the Hanging Gardens of Babylon. The Hebrews also knew it, since in the Old Testament in the Book of Exodus, "*manna*" is described as being like white coriander seed. In the third century BC, Romans also found coriander to be an excellent seasoning for popular food dishes. Besides, fresh leaves of coriander are now extensively used for garnishing many Latin American, Russian, East European, Chinese and Indian food dishes. The Spaniards call fresh coriander leaves "*cilantro*" and it is also called "Chinese Parsley" as it belongs to the parsley family, Umbelliferae. Coriader is a naïve of the Mediterranean region.

K. P. Nair, *Minor Spices and Condiments*,
https://doi.org/10.1007/978-3-030-82246-0_7

Commercial Production

Coriander is commercially grown in India, Morocco, Russia, Hungary, Poland, Romania, Czechoslovakia, Bulgaria, France, Guatemala, Mexico, Argentina and the USA. In the late twentieth century, India had about 400,000 ha with an average production of 180,000 tonnes. Coriander is grown in most of the Indian states and is an important subsidiary crop grown in black cotton soils of the Deccan plateau in Andhra Pradesh, Maharashtra, and, Tamil Nadu, where it grown as a rainfed crop. In the rich silt loams of northern India (Rajasthan, Uttar Pradesh and Madhya Pradesh) also in western India (Haryana) coriander is grown as an irrigated crop.

Botanical Description

Coriander is a small, thin-stemmed, bushy herb, about 25–30 cm in height, with many branches, tops and umbels. Leaves are alternate, compound and petiole has a pair of stipules, which sheath the stem at the base. The compound leaves become highly segmented and linear as they reach the upper extremities. Both leaves and stems, nay, the whole plant, has a strong pleasant aroma for which it is liked and used in cookery for flavoring and garnishing the dishes. The inflorescence is a compound umbel and usually comprises about five smaller umbels. The corolla is made of five petals. Flowers are both bisexual and unisexual, the latter being mostly staminate. The flower possesses five stamens and the ovary is inferior and bilocular. The fruit is globular, round and small (3–4 cm in diameter). Its surface has longitudinal fine grooves and ribs alternately. When pressed the fruit easily breaks into two locules, convex outside ad concave inside, in each of which there is one seed. The fruits are mild, delicately fragrant due to the presence of volatile oil (0.2–1.0%) in them, according to the variety, in which region the plant grows. In cultivated fields, two distinct varieties can be seen, one having light-green stem and the other with a somewhat light purple blush on it. The former plants have purely white flowers, while the latter have highly purple colored flowers. Seeds are pale white to light brown in color.

Economic Importance and International Trade

Coriander seed is one of the oldest and most important condiments consumed all over the world. Additionally, its fresh leaves and tender stems are extensively used in India and elsewhere to garnishing/flavoring curried dishes of all sorts and the Indian "chutney" (a side dish eaten during the widely popular South Indian *dosa*) making, pickling, manufacture of curry powder (up to 40% used), and also in the Indian Ayurvedic medicine. During 1995–96 India recorded an all-time high export of coriander. USA imports the maximum of coriander from India. Interestingly, the maximum percentage of export of 68% was from Canada. Heavy black cotton soils are preferred and soils should be supplemented with heavy doses of organic manure.

Land Preparation Since the plant grows best in soils with good tilth, they should first be plowed by a soil turning mold board plough, 2–3 ploughings, followed by a *desi* plough. To prevent moisture loss, the field must be planked immediately after ploughing.

Fertilizer Application Fields are first dressed with 10–15 tonnes of well rotten farm yard manure supplemented with N, P, and K in the ratio of 20 kg–30 kg- and 20 kg. In the case of irrigated crop, the entire P and K are applied at the time of planting and one half of N and the second half is applied in two split doses at the time of first irrigation and the second dose at the time of flowering.

All the three nutrients, N, P and K, had significant effects on the plant height, number of branches, leaf number and fresh weight of coriander plants. 80 kg N/ha produced better plant height, leaf number and fresh weight and at 40 kg N/ha the plant had more primary and secondary branches. P had a better effect on plant height at 60 kg/ha than at 30 kg/ha. K at 30 kg/ha was optimum for all growth parameters (Bhat et al. 1992) (Table 1).

Increasing Seed Germination Soaking coriander seeds for 12 h in growth regulators, GA (Gibberlic Acid) 50 ppm solution/NAA (Naphthelene Acetic Acid) 20 ppm solution/IAA (Indole Acetic Acid) 10 ppm solution enhances germination remarkably. Maximum gemination was observed with GA treatment in gunny bag cloth rapping followed by NAA treatments (Badgujar and Wauhal 1990).

Irrigation 3–4 irrigations are given for irrigated crop, first after 35–40 days of planting, second 60–70 days after and the third 80–90 days after and fourth 100–110 days after.

Weeding and Hoeing First weeding and hoeing done after 40–45 after planting. Second hoeing should be done before the rows close up. For a good crop, weeding and hoeing are necessary to keep the fields weed free and also to conserve soil moisture.

Plant Protection Measures

Disease Control Wilt, powdery mildew and stem gall are here main diseases of coriander. Wilt caused by *Fusarium oxysporum* var. coriander can be controlled by the following measures

1. Use healthy disease free seeds, obtained from wilt free precious crop

Table 1 Latest improved varieties of Coriander developed by the National Research Center for Spices (which later became the Indian Institute of Spices Research in Calicut, under the administrative control of the Indian Council of Agricultural Research in New Delhi) in collaboration with State Agricultural Universities

Variety (Year developed)	Parentage	Special characters	Crop duration	Yield(kg/ha)	Where developed
Gujarat Coriander1(1974)	Local selection	High yield	112 days	1100	Jagudan(Gujarat Agrl Univ)
Col (1981)	Germplasm selection	Tall plant, many umbels	110 days	500	Coimbatore (Tamil Nadu Agrl university)
Co2(1982)	Reselection	High yield	90–100 days	600–700	Same as above
Gujarat coriander 2(1985)	Selection	High yield	100–115 days	1500	Jagudan (Gujarat Agr Univ)
RCr41 (1988)	Recurrent selection	High yield	130–140 days	1200	Jobner(Rajendra Agr Univ)
Rajendra Swati (1988)	Mass selection	High yield	100 days	1200–1400	Dholi (Rajendra Agr Univ)
Swati (1989)	Mass selection	High yield	80–90 days	885	Guntur(Andhra Pradesh Agr Unv)
Sudhana (1989)	Mass selection	High yield	95–105 days	1025	Same as above
Co3 (1991)	Pure-line selection	High yield	90 days	640	Coimbatore (TNAU)
CS287	Recurrent selection	Small stature	78 days	510	Coimbatore (TNAU)
Lam selection	Mass selection	Medium, bushy	110 days	1500	Guntur(Andhra PradeshAgr Univ)
CS2 UD 20	Not available	Medium, tall plant	100–110 days	1175	Jobner(Rajendra Agr Univ)

Note: National (Indian) average yield: 470 kg/ha, Seed rate: 10-12 kg/ha, Planting season: *Kharif*-June-July, *Rabi*-October-November, Fertilizer Application: 10–15 tonnes Farm Yard Manure/ha (FYM) (basal application and N-P-K in the ratio of 30–40-20 kg/ha for irrigated crop; 25 tonnes FYM/ha (basal application and N-P-K in the ratio 15–20-0 kg/ha for rainfed crop. Source: Indian Institute of Spices Research, Calicut, Kerala State, India

2. Treat seeds with mercurial fungicides, Agrosan GN, Ceresan or Hexasan @2 g/kg seed
3. Do not grow coriander during 3 years in a wilt infected field
4. Deep summer ploughing in summer will also help control the disease to an extent

Controlling Powdery mildew caused by *Erysiphe polygoni* or *E.heraclai*: The disease can be effectively controlled by spraying the crop with wettable sulfur @ 1 kg/ha, dusting the crop with 20–25 kg/ha sulfur or spraying Cosan @ 100 g/100 l water. Dusting may be repeated after a fortnight, if found necessary.

Controlling Stem gall caused by *Termonas (Protonyce mcrosporum)* cannot be controlled by spraying/dusting any fungicide. Variety Karan (UD 41) is found to be highly resistant and it is advisable to treat seeds with Thiram before sowing.

Controlling Insect Pests

1. Aphids: Coriander is normally attacked by aphids after flowering. Though aphids can be controlled by a number of insecticides, it is not advisable to follow this mode, because, the bees which facilitate pollination would be severely affected by insecticides. As a restricted recommendation, spraying 0.03% Rogor solution may be resorted to. Spraying should be done in the evening when the bee population is minimal, and, it is definitely preferable to complete spraying before flowering onset.
2. Leaf eating caterpillars: These attack the coriander plant at the seedling stage. They can be controlled by spraying 0.03% of folidol.

Additional Plant Protection Measures to Control Diseases and Pests

1. Grain mold can be checked by spraying carbendazim (0.1%) 20 days after flowering
2. To control powdery mildew two sprays of karathane (0.1%), first at the time of disease initiation and second 10 days later
3. Root rot is controlled by drenching in bavastin (0.1%) twice at 10–15 days interval
4. Wilt incidence can be controlled by seed pelleting with *Trichoderma viride* and neem cake application @150 kg/ha
5. Aphids can be controlled by spraying endosulfan (0.07%) twice, first at flower initiation and second a fortnight later
6. Both aphids and red spidermites can be controlled by two sprays of endosulfan (0.1%) at fortnightly intervals

Protection from Frost Frost damage is severe in Rajasthan and Madhya Pradesh states of India especially when the crop is grown rainfed. In irrigated area an irrigation when the frost is expected will save the crop while in rainfed area fumigation at dawn when temperature is lowest can only save the crop.

Tissue Culture The manipulation of culture media, and the utilization of fresh leaves of seedlings and hypocotyl segments in in coriander and cumin resulted in callus, embryos and adventitious shoots (Jhan et al 1982).

Harvesting and Threshing

Coriander crop ripens in 115–125 days depending on the variety and agro climate. Crop must not be allowed to over ripen as this will adversely affect the quality of the produce, especially the color. Crop must be harvested when 60% of seeds in the main umbels turn yellowish brown. Harvesting must be completed early in the morning when there is dew, by uprooting the plants or cutting with sickles, bundled and kept upside down in shade so that the grains are not directly exposed to the sun and they lose the green color. When the plant bundles are sufficiently dry, the grains are separated either by beating the bundles with wooden sticks or trampled by feet to separate them from the stalks. The grains are winnowed to separate chaff and other materials, kept stored in moisture free rooms until sale.

Yield Irrigated crop yields vary from 10–15 q/ha, rainfed one 4–5 q/ha, when mixed with other crops it will be 1.5–3 q/ha (q = quintal = 100 kg). If the recommended cultivation package is followed, irrigated crop can yield up to 25 q/ha, and, rainfed one 20 q/ha. Srirama Rao et al. (1981), while investigating the magnitude of genetic variability in 42 varieties of coriander, concluded that traits, such as, (1) plant height, (2) number of umbels and (3) weight or density of grain contributed much to the yield, and, the coriander breeders must give ample weightage to these traits, in their breeding programs (Table 2).

The European Spice Association (ESA) mandates the following specifications for coriander:

Post Harvest Technology

Preparation for the market: The seeds are winnowed and cleaned of dust, stones, leaves, stalks and any other extraneous matter. They are ready then for packing and marketing.

Cleaning and Destoning of Coriander Seeds: A Simple Technique

A dry process was developed for the separation of stones and sand particles from coriander seeds, using principles of fluidization with air as a medium. The method works on the principle of minimum fluidization velocity differences between stones and coriander seeds. When the mixture is fluidized, the particles with the minimum fluidization velocity but higher than that of the coriander seeds settle down at the

Table 2 ESA specification for coriander

Property	Specification
Total ash (%w/w max)	7(ISO)
Acid soluble ash (% w/w max)	1.5(ISO)
H2O (%w/w max)	12 (ISO)
Volatile oil (%w/w min)	0.3 (ESA)

Note: ISO = International Organization for Standardization

bottom of the column. Those particles with a minimum fluidization velocity less than that of the coriander seeds get mixed up with the seeds, and these fine particles can be removed by sieving.

Quality Grading

The grade designations are "Special", "Good", "Fair" and "N.S.", mandated by both Agmark (Indian) and ISI (British Standards Institution) (Table 3).

Urgent Need for Development of Better Coriander Varieties

Indian coriander varieties are generally not preferred for distillation to get essential oil in Western countries, because, these Indian varieties are low in essential oil content and are non-uniform in color and grade. Hence there is an urgent need to develop high yielding, firm. (like the Russian varieties which are not easily breakable), disease resistant varieties rich in essential oil and adaptable to various agro climatic regions of the country. This is an urgent need not only to tap better the export potential, but, also to maintain high internal supply. By increasing Indian production, maintaining seed quality, and enlarging production base of by products, and thereby keeping the price competitive, India could play a major role in international coriander market.

Table 3 Composition of Coriander Seeds and Coriander Leaves

Property	Coriander Seed	Coriander leaf (Fresh)
Moisture	6.3%	87.9%
Protein	1.3%	3.3%
Volatile oil	0.3%	
Non-volatile ether extract	22%	
Total ether extract (fat)	19.6%	
Crude fiber	31.5%	
Carbohydrates	24%	
Mineral matter (total ash)	5.3%	1.7%
Acid insoluble ash	0.2%	
Calcium	0.8%	0.14%
Iron	0.006%	0.01%
Phosphorus	0.44%	0.06%
Sodium	0.02%	
Potassium	1.2%	
Vitamins (mg/100 g)		
Vitamin B1	0.26	
Vitamin B2	0.23	60
Niacin	3.2	0.8
Vitamin C (Ascorbic acid)	12.0	135
Vitamin A	175 IU/100 g	10,640 IU/100 mg

Packaging and Storage

Whole Coriander The cleaned and well dried seeds can be stored in gunny bags with paper or polythene lining to prevent loss of volatile oil. For seed purposes, it is preferred to store grains in cloth bags or gunny bags to avoid loss of germination of the seed.

Ground Coriander In commerce, coriander powder is generally offered for sale in quantities of 100 g pre packings in paper and polythene pouches. Often these packages do not provide the requisite protection to the product against losses of volatiles, seepage of fat and ingress of moisture during storage. In recent years, flexible plastic films and foil laminates with better functional physico-chemical properties have made inroads in packaging of this product in consumer packages to provide necessary protection.

Humidity-moisture relationship investigations on ground coriander have revealed that a moisture content above 12% (dry weight basis) is critical in respect of storage stability of the product. Aluminum foil laminate is found to offer maximum protection against volatile oil loss, seepage of fixed oils, and ingress of moisture. Double pouch of MSAT cellophane and low density polythene (LDPE), MXXT cellophane pouch and polycell laminate pouch are equally good in offering desired protection to the product. Single printed polythene pouches are found inadequate in preventing loss of volatiles and seepage of fixed oils during storage, and printing on polythene gets smudged, especially under accelerated storage conditions.

Storage of Seed Coriander seeds should be stored under cool, dark and dry conditions in order to prevent (1) browning, (2) loss of flavor and (3) ingress of moisture. Cilantro (dried or freeze-dried leaves) should be further protected from flavor loss by making certain that container is always tightly closed after every use. Coriander seeds can also be cold stored by packing them in sound hessian bags internally lined with polythene as a protection against ingress of moisture under humid conditions. Under good storage conditions, coriander seed will retain flavor and color for about 6–9 months.

Natural Occurrence of Aflatoxins in Coriander Seeds

Secretion of some toxic metabolites by certain fungi during food storage has created an alarming situation as they are consumed by human beings. *Aspergillus* sp., *Penicillium* sp., and *Fusarium* sp., secrete aflatoxins on coriander seeds stored under unhygienic conditions and dampness. These secretions occurred during 6 months of storage. Aflatoxins were isolated from 10 seed samples of *Coriandrum sativus* Linn. Of these, only five were found to be BGY positive; aflatoxins B1, B2 and G1 (1, 2, 1 substrate please) were isolated from BGY positive samples. The concentration of various aflatoxins was determined after development, with chloroform - acetone from observation on their flourescene emission on silica gel plates (Misra 1987).

Processed Products

Volatile Oil It varies from 0.1% to 1.7%, depending on seed size, region where the plant is grown, soil and climate. Indian coriander oil is colorless, pale-yellow liquid, with a characteristic odor and taste and has the following characteristics: specific gravity at 15 ° centigrade: 0.8715–0.8760, refractive index at 25 ° centigrade: 1.4569–14,612; optical rotation at 25 ° centigrade: +10 0(last 0 superscript please) +13 0(0 superscript please).

The volatile oil is made up of hydrocarbons and oxygenated compounds. The former accounts for 20% of the essential oil. The major oxygenated compounds present are: d-linalool or coriandrol (45–70%). The oil causes irritation to skin when it is in contact for a long time.

The unripe fruits and other parts of the plant yield small amount of an inferior oil with a bug-like odor, which, however, disappears on keeping, evidently due to the polymerization of the odoriferous principles.

Adulteration of Volatile Oil Commercial volatile oil of coriander is extensively adulterated with sweet orange oil, turpentine and ethanol or aniseed oil or even cedar- wood oil. In flavoring, coriander volatile oil has the advantage of being more stable and retaining its agreeable odor longer than most other oils of its class.

Fatty Oil (Non-Volatile Oil) Seeds contain 19–21% of fixed fatty oil, having a dark brownish-green color and odor similar to that of coriander oil. It has the following characteristics; specific gravity at 15 ° centigrade: 0.9262–0.9284; refractive index at 30 ° centigrade: 1.4704; saponification value: 182–190; iodine value: 93–100; unsaponified matter: 2.3%. The components of insoluble fatty acids are: palmitic: 8%; patrosslinic: 53%; oleic: 32% and linoleic: 7%. The oil solidifies on keeping. Sodium soap prepared from the oil has pleasant odor and good leathering properties. It is soft in consistency and green in color.

Coriander Herb Oil Stalks and leaves of coriander yield considerable volatile oil (about 0.1–0.95%), but, it is absolutely impossible to use the leaf oil instead of seed oil in any flavor, perfume, or pharmaceutical formulae. The leaf or herb or herb volatile oil has a pronounced odor of decylaldehyde or other higher fatty aldehyde.

Coriander Oleoresin Coriander has more fatty oil than volatile oil, and the former interferes in the extraction of the oleoresin. It is desirable to lower the fatty oil and enhance volatile oil suitably. Volatile oil content of commercial oleoresins ranges from 5 to 40%. The former USSR has succeeded in the preparation of coriander oleoresin. India has entered the world market and is making an impact.

Coriander Dal – An Improved CFTRI Product

Processing of coriander into dal (Lentils and Pulses) cotyledons has been made easier by a CFTRI technique which gives a higher yield than the traditional method. The traditional method of Dhania (Coriander) Dal is to subject the coriander seed to overnight soaking in saline water, partially dry it, and hand pound to remove the husk. This method is laborious and time consuming, giving only up to20% dal constituents and 20% brokens. The CFTRI process utilizes indigenous milling equipment, and is based mainly on an improved conditioning technique. The pilot plant tried out by CFTRI has a capacity to process 60 kg seed/hour, giving an output of 40% whole polished dal, 10% broken dal and the rest 50% husk, bran, and germ. The byproducts generated by this process, namely, husk, bran and brokens, can be used in the preparation of spice mixes. The estimated investment on equipment and machinery is about INR 200,000 (INR 75 = U1 US $).

Dhania Dal Principal use: This is mainly used in Gujarat and some north Indian states as an adjunct to *supari* (the arecanut chew) to make which the coriander seed is first dehusked, flaked and then given a mild heat treatment and slated. The treated seeds may also be eaten as a highly flavored digestive chew.

The traditional method of making Dhania Dal is to subject the coriander seed to overnight soaking in saline water, partially drying it, and hand pounding to remove the husk. This method is laborious and time consuming, giving an output of 20% dal and 20% brokens.

Seasoning from Coriander Roots

In Thailand, they derive, yet, a third "seasoning" from coriander roots which are claimed to be as distinctly different in flavor from seeds, as laves are from seeds.

In the USA, the largest importer of spices, five distinct products of coriander are in the market:

1. Dried and freeze-dried green leaves of coriander
2. Dried coriander seeds (whole and ground)
3. Coriander essential oil
4. Coriander oleoresin
5. Thai "Coriander Root Seasoning" for foods

Uses of Coriander

As Flavorant The fresh green stems, leaves and fruits of coriander have a pleasant aromatic odor. The entire plant, when young, is used in preparing chutneys and sauces, and the leaves are used for flavoring and garnishing curries and soups. The fruits (seeds) are extensively used as a condiment with or without roasting in the preparation of many foods, oriental, European and North American. Possibly, coriander is the first spice used by mankind, as early as 5000 BC.

In Medicine Extensive use in human health, as a carminative, diuretic, tonic, stomachic, antibilious, refrigerant and aphrodisiac. The volatile oil from coriander seed is used as a flavoring agent for spiritous liquors and in the cocoa and chocolate industries.

In Perfumery Oil of coriander seeds is a valuable ingredient in perfumes. It's soft, pleasant, slightly spicy note, blends into scents of oriental character. It harmonizes well with jasmine, imparting life and lift to otherwise dull or too synthetic compositions. Decyladehyde (0.1% volatile oil), obtained by treating volatile oil of coriander with bisulfite, is a perfumery.

Oleoresin and its additional uses: Good quality oleoresin extracted from coriander seeds, can be used to flavor beverages, pickles, sweets, sausages, and other delicacies, such as snacks. The residue from distillation can be used as a good cattle feed, as it is rich in protein and fat.

Soluble Coriander (Superesin) This is prepared by proper blending and dispersing a minimum 3% of total extractives of coriander on a soluble, dry edible carrier. The coriander extractives are the sum of the non-volatile ether extract and volatile oil content. The coriander flavoring consists of a maximum of 33% of volatile oil of coriander (vol/wt), the remainder as the non volatile ether extract of coriander. The coriander extractives are of good quality and possesses the characteristic flavor of good grade coriander. Their flavor is unimpaired in or during the process of extraction.

References

Badgujar CD, Wauhal KN (1990) Effect of soaking and wrapping of seeds on germination of coriander. Spice India 3(10):9–11

Bhat VR, Sulikeri GS, Hulamani NC (1992) Coriander: effect of NPK on herbage yield of coriander. Spice India 5(10):11–13

Misra N (1987) Mycotoxins in spices. Part III. Investigation on the natural occurrence of aflatoxins in *Coriandrum sativum* L. J Fd Sci Technol 24(6):324–335

Srirama Rao T, Babu MK, Bavaji JN (1981) Path-coefficient analysis of seed yield in coriander. Indian J Agric Sci 51(10):76–78

Cumin Seed or Safaid Jeera

Botanical Name: *Cuminum cyminum* Linn.
Family: Umbelliferae

Description and Distribution

Cumin or *Safaid Jeera* comprises the dried pale-yellowish to light greyish-brown seeds of a small slender annual herb of the coriander family, believed to be native of Egypt, Syria, Turkestan and the East Mediterranean region. It grows to a height of 30–45 cm and produces a stem with many long branches, finely divided with deep green leaves and small flowers, white or rose in color, borne in umbels.

The aromatic seed-like fruit, commonly known as 'seed", is elongated, oval, elliptical, deeply furrowed, approximately 6-mm long, and light yellowish brown in color, somewhat similar to caraway seed, but slightly loner and not curved in shape. The odor is peculiar, strong and heavy, pleasant to some, but, disagreeable to many others, while flavor is warm, slightly bitter and somewhat disagreeable.

Brief History and Distribution

Cumin is one of the oldest spices known since Biblical times. To day, the plant is grown extensively in Iran, India, Morocco, China, southern Russia, southern Europe, UAR and Turkey. Iran is the major world exporter of cumin seed known as "green cumin", and the country produces close to 50,000 tonnes, based on weather conditions. However, India ranks number one in the world in terms of area and production.

Major types of cumin seeds in International trade:

K. P. Nair, *Minor Spices and Condiments*,
https://doi.org/10.1007/978-3-030-82246-0_8

1. Iranian Cumin: Pale green in color and has about 2.5–3.5% essential oil. It is machine cleaned product marketed as Iranian cumin without grade distinction. Iran also has a wild herb called by natives as black cumin.
2. Indian cumin: India produces the most cumin in the world and also consumes the most. Quite similar to Iranian cumin in flavor and aroma. Indian cumin is golden brown in color and the essential oil is of prime quality, ranging between 3 and 5%.
3. Egyptian, Turkish Cumin: Essential oil content varies from 1.5–5%, different in flavor from Indian and Iranian cumin.
4. There are also cumins from other regions of the world, varying in essential oil content from 1.5–5%, with some exceptions having up to 6%.

Area and Production

In India cumin is cultivated in almost all states except in Assam, Bengal and Kerala, notably, in Rajasthan, Gujarat, Uttar Pradesh, Andhra Pradesh ad Tamil Nadu.

International trade: India hardly exports 4% of its total cumin production, but, it is the largest exporter of this spice. India exports cumin seeds to more than 50 countries and has established markets in USA, UK, Kuwait, Japan, Singapore, Nepal and Bangladesh. But, the country has to compete with Turkey, Iran, Syria and Lebanon.

Crop Management

Climate and Soil Cumin is a tropical plant and it can be cultivated in *Rabi* in any Indian region where the atmospheric humidity during February–March, when the plant is flowering, and seed formation takes place, is low. High humidity during flowering ad fruiting period would induce disease infestation.

Land Preparation Land must be thoroughly ploughed to obtain a fine soil tilth, first by a soil turning plough, followed by 2–3 ploughings with a *desi* plough. Field is then planked and seed beds (2 m × 3.5 m) are made to facilitate irrigation, and liberal application of FYM is made.

Manures and Fertilizers No additional fertilizers or manures are required if the preceding crop has been fertilized with 15–20 tonnes of FYM. If not, 15–20 tonnes/ha of FYM should be applied at the time of planting. Additionally, a single application of urea (25–30 kg N/ha) 45–50 days after sowing with irrigation must be made.

Weeding *Zeeri* (Hindi – Indian language) is the most important weed of cumin. Commercial cumin may contain this weed seeds if not controlled efficiently in the field. As the weed is distinguishable at the flowering stage, hand weeding is followed at this stage, to uproot the weeds. Other weeds like *Chenapodium album*, *Chenapodium maturalium*, *Cypeus* sp., etc. are the other less severe weeds which can be controlled by hand weeding or a spray of 2, 4-D or 2,4- 5 T.

Cultivars

There are four cultivars of cumin, as follows:

1. Tall
2. Dwarf
3. Pink-flowered
4. White-flowered

The pink-flowered variety yields more than the white-colored ones (Narayana Swamy and Patil 1984) (Tables 1 and 2).

Table 1 Improved varieties released by IISR

Variety & Year released	Parentage	Special characteristcs	Duration	Yield	Center where developed
MC 43 (1970)	Selection from local Germplasm "Mehsana local"	Yellow stem, bold grain, high yield, 3.6%oil, tolerant to wilt & blight diseases	115 days	560(kg/ha)	Jagudan (Gujarat Agri univ)
Gujarat Cumin 1 (1983)	Pure line selection	High yield, bushy tolerant to wilt & blight	105 days	735(kg/ha)	Same as above
RZ 19 (1988)	Recurrent selection from "Kakri local" "UC19"	High yield, erect, tolerant to wilt & blight	125 days	590(kg/ha)	Jobner(Rajendra Agri.Univ)

National average yield: 520 kg/ha; Seed rate: 10-20 kg/ha Planting season: November; NPK(kg/ha):30:15:0 Irrigation: 3–5.

Table 2 Varieties recommended for different Indian states by IISR

Cumin variety released	Yield (kg/ha)	Indian State recommended for cultivation
S 404	350	Gujarat
MC 43	580	Gujarat
Guj. Cumin 1	700	Gujarat and Rajasthan
RZ 19	500	Rajasthan and Gujarat
Cuj. Cumin 2	700	Gujarat

Sowing Time

In Gujarat and Rajasthan states of India sowing is done between November 15 and December. Good yield is obtained through early sowing in November and late sowing results in poor yield. Normally the high ambient temperature before November 15 is not conducive to good germination and establishment of the cumin plant.

Seed Rate A seed rate of 12–15 kg/ha of good quality seed is sufficient to obtain an optimum plant stand and yield as well.

Sowing Method It is a wrong practice to first sow the seeds, by broadcasting in the fields, and then make seed beds. The proper method would be to first make the seed beds and then sow the seeds and cover them up in the seed bed with an iron rake. Care should be taken to ensure that the seeds are not sown below 1.5 cm–2.0 cm deep, with a spacing of 30 cm × 15 cm. Seeds after swing should be covered with a thin layer of soil. To prevent fungal attack, it would be advisable to treat seeds with a mercurial fungicide (agrosan or ceresan) at 2 g/kg of seed or treat the seeds with N-trichloromethyl at the rate of 4 g/kg seed.

Irrigation A light irrigation should be applied immediately after sowing. A second irrigation is required after about 8–10 days. Seeds will only germinate after second irrigation. A third irrigation is given after about 5–6 days, which will help seedling emergence and its firm establishment in the soil. After this, the crop is irrigated at an interval of 12–20 days depending upon the soil and weather conditions. Irrigation should be avoided when the crop is maturing, because this will adversely affect seed quality. The last irrigation at the time of grain maturity should berather copious so that additional irrigation is not required during grain maturity.

Interculture (Hoeing, Thinning and Weeding) First hoeing and weeding is done when the plants are about 5 cm tall, in order to facilitate further crop growth. At this time thinning is also done to maintain plant to plant spacing of 15 cm, following this, 1–2 hoeings are done (Table 3).

Commercial Quality

From the commercial quality view point cumin seeds should have the following composition (Table 4).

The exhausted cumin seed powder had 0.2–1.2% volatile oil content and very low cold water extract (3.2–4.6%). Such exhausted cumin seeds find a way into adulteration into the genuine cumin seeds, like in the case of coriander, by unscrupulous traders, which should be severely penalized by the governmental authority. If not, the consumer stands totally cheated.

Table 3 The composition of cumin seed

Property	Content
Moisture	6.2%
Protein	17.7%
Fat	23.8%
Crude Fiber	9.1%
Carbohydrates	35.5%
Mineral matter	7.7%
Calcium	0.9%
Phosphorus	0.45%
Iron	0.048%
Sodium	0.16%
Potassium	2.1%
Vitamins mg/100 g	
Vit B1	0.73%
Vit B2	0.38%
Niacin	2.5
Vit C	17.2
Vit A	175 IU/100 g
Calorific value (food energy)	460 calories/100 g

Cumin Powder

Cumin powder is gradually gaining commercial importance in world trade since 1992–93. It is a good value added product like the oleoresin and essential oil with much potential in world trade.

Volatile Oil

The dried fruits are crushed and immediately thereafter, steam distilled to yield 2.5–4.5% of volatile oil, colorless or pale yellow, turning dark on keeping. The yield of the oil depends upon the quality and age of the seed – the older seeds contain less oil. The range of variations in physico-chemical properties of the oil produced in India and other countries is summarized as follows

1. Specific gravity at 25 ° centigrade: 0.8923–0.9246 (average 0.8969)
2. Optical rotation: +3 + 6 o(o superscript please)30o(o superscript please) (average 3o(superscript please)56o (o superscript please)
3. Refractive Index at 20 ° centigrade: 1.4949–1.5060 (average 1.4968)
4. Aldehyde content: 32–42% (average 36.30%)

   The chief constituent of the oil is cuminaldehyde (20%–40%) which is used in perfumery.

Table 4 Commercial quality composition of cumin seed

Property	Content
Moisture	6–10% (average 7.7%)
Volatile oil content	2.5–3.6%(average 3.1%)
Cold water extract	14.9–20.4% (average 17.5%)
Total ash	6.9–7.6% (average 7.3%)
Acid insoluble ash	0.2–0.8% (average 0.4%)

New Supercritical Fluid Extraction (SCFE) Process for Volatile Oil

A new extraction process for isolation of essential oil/extractives from natural substances using liquid and dense carbon di oxide has been described. Cardamom (fruits), cumin (seeds), clove (buds), ginger (rhizomes), parsley (seeds), mace (arils), sandalwood (stems) and vetiver (roots) are used as extraction materials in a specially designed high pressure Soxhlet apparatus for preliminary investigations of liquid carbon di oxide extraction. The essential oil/extractives obtained by this method are found superior in quality and flavor as compared with conventional steam-distilled essential oil. The process conditions for extraction of clove, parsley and vetiver with liquid and dense carbon di oxide investigated at different pressures and temperatures in a high pressure gas extraction are found at subcritical state (Pruthi 1996).

Adulteration of Cumin Volatile Oil The most troublesome adulterant is synthetic cuminaldehyde, the presence of which cannot be detected analytically except that this adulteration will show a rise in optical rotation value.

Fixed Oil In addition to volatile oil, cumin seed will contain about 10% fixed oil (non-volatile), greenish-brown colored oil with a strong aromatic flavor. It is a semi-drying oil having an iodine value of 92.

Cumin Oleoresin Since 1986–87, cumin seed oleoresin is also being manufactured and exported from India. The product has tremendous potential as an export commodity.

Uses

As a Flavorant Cumin seeds have an aromatic odor and a spicy and somewhat bitter taste. They are largely used as a condiment and form an essential ingredient in all mixed spices and curry powder to flavor soups, pickles and for seasoning breads and cakes. They are also candied.

In Holland and Switzerland, cumin is used to season some kinds of cheese, whereas in Germany and Switzerland, it is used to flavor bread, cakes and pastries. Cumin makes an excellent seasoning material for soups and stew and is widely used for this purpose throughout Latin America. It is also used in preparing dishes in Central and South America. On account of the growing popularity of cumin, the demand for Indian and Iranian cumin is on the rise and it will be advantageous to both countries to tap this great potential.

Most of the cumin used in USA goes into chilli powder, followed by manufacture of "chilli concame", and "chilli meat products"and Mexican-style foods in which it is constantly used. However, it is also an excellent spice for rice dishes, curries, stuffings, sauces and marinades, in particular for Middle Eastern dishes. After chilli and peppers, it is cumin which is the most widely used spice to flavor Latin American foods, in general.

Use in Medicine Cumin seeds have for long been considered as a stimulant, carminative, stomachic, astringent and useful in diarrhoea ad dyspepsia. They are now widely used in veterinary medicine. An infusion of cumin relieves flatulence.

Cumin Seed Oil Used in flavoring compounds, specially curries and culinary preparations of oriental character. Because of its powerful odor and flavor, it must only be used sparingly and carefully. It is used in perfumery and to flavor liqueurs and cordials.

Fixed Oil The residue left after extraction of volatile oil contains 17.2% protein and 30% aromatic greenish fixed oil or fat. Its cake can be used as cattle feed. Also it can find use in oil, fat and soap industries.

References

Narayana Swamy, Patil A (1984) Agro-techniques for cumin cultivation. Ind Cocoa, Arecanut and Spice J 8(1):10–11

Pruthi JS (1996) A novel, efficient, and versatile technique for spice extraction. Proceedings of Asian Symposium on Super-Critical Fluid Extraction Technology of Natural Products, New Delhi, India. pp. 101–109

Fenugreek

Botanical name: *Trigonella foenum-graecum*.Linn

Description and Distribution

Fenugreek is one of the oldest cultivated plants. Medical Papyri from ancient Egyptian tombs reports that fenugreek was used in those days to reduce fevers and as a food. In religious rites, fenugreek was one among the components of Holy Smoke, an Egyptian incense, used in fumigation and embalming. In the second century BC, Roman history states that Porcius Cato (the then Roman Authority in Animal Husbandry) ordered that fenugreek should be sown as a fodder crop for oxen in the Roman fields. In the Middle Ages, fenugreek was recommended as a cure for baldness in men.

Fenugreek is a spice used in India since 3000 years. It belongs to the "Pea" family, is popularly known as *Methe* (in the popular Indian language, Hindi), and its leaves, both fresh and dried, known as *Methee* or *Methi*. Seed is the dried ripe fruit of an annual herb, native to south-eastern Europe and West Asia, now cultivated in India, Argentina, Egypt and the Mediterranean countries (southern France, Morocco, and Lebanon). The seed is small and yellowish-brown in color. It has a pleasantly bitter taste and a peculiar odor and flavor of its own, very much used in making curry powder, pickles and other spice blends.

The seed is produced as a spice, leaves and tender pods constitute as one of the most popular leafy vegetables for human consumption, and also as a fodder for cattle. To some extent, seed is also used as a medicine. The robust herb has light green leaves, is 30–60 cm tall, and produces slender beaked or sickle-shaped pods, 10–15 cm long. Each pod contains 10–20 small hard yellowish brown colored seeds, which are smooth and oblong, about 3 cm long and grooved across one corner, giving them a hooked appearance.

K. P. Nair, *Minor Spices and Condiments*,
https://doi.org/10.1007/978-3-030-82246-0_9

Area, Production and Export

India is one of the major producers and exporters of fenugreek, grown on an extensive scale in states like Rajasthan, Madhya Pradesh, Uttar Pradesh, Gujarat, Maharashtra, and Punjab. The spice occupies the third place in area and fourth in production among all the minor spices grown in India. India exports only about 10% of its production. Export potential to Middle East is high because of the proximity of the region. Both fenugreek powder and oleoresin have good export potential. Main markets are Japan, Singapore, Saudi Arabia, Democratic Republic of Korea, Kuwait, Nepal, Yemen, all together constituting 70% of the Indian export. Lately, India is exporting substantial quantity of fenugreek oleoresin.

Crop Management

Climate and Soil

Fenugreek is a cold season (*Rabi)* crop and fairly tolerant to frost and low temperature. The plant can be pot-cultured throughout the year under proper irrigation. Crop grows on all kinds of soils, does best in loamy soil. Can also be grown in heavy black cotton soils where a second crop can also be taken.

Land Preparation For good early crop growth and higher yield, the land should be well prepared. The soil should be ploughed 4–5 times first with a soil turning mold board plough followed by a *desi* (indigenous) plough 3–4 times. The field is then well levelled by planking, and, seed beds are made depending upon the method of irrigation, about 3–5 m long, and 1.25–1.75 m wide.

Manuring

The soil should be well manured to produce succulent leafy matter and early growth. Application of well rotted FYM @360quintals/ha is recommended before ploughing. For seed crop, 2.5 q bone meal or 2 q of superphosphate should be applied before ploughing. Top dressing of 1 q/ha of ammonium sulfate after the first cutting reduces excessive leaf growth. Alternatively, three top dressings, each of about 45 kg of ammonium sulfate/ha the first 4 weeks after sowing, the next two after every two cuttings.

As per recent research at IARI, New Delhi, India, two foliar sprays of urea (1%) about 45 and 60 days after sowing gives better seed and green leaf yield. One can take 2 leaf cuttings before allowing the crop for seed harvest (Mishra 1987) (Table 1).

Table 1 New Fenugreek varieties released by IISR

Variety & Year released	Parentage	Special characteristics	Duration	Yield(kg/ha)	Where developed
Col (1982)	Reselection: TG 236(introduced by IARI)	Dual purpose, quick growing, high seed protein	90 days	Grain: 685 Greens: 4–5 tonnes	Tamil Nadu Agri Univ
Rajendra Kanti (1988)	Mass selection Germplasm: RM 16	High yield, bushy resistant to leaf spot	120 days	1200–1400	Rajendra Agri Uni
RMt 1 (1989)	Pure line selection Nagpur local "*Methi*" (NLM) "Prabha"	Tall, high yield tolerant to root rot and powdery mildew	145 days	1560	Rajendra Agri Univ
Lam Sel.1	Selection: Germplasm	High yield, bushy	68 days	740	Andhra Pradesh Agri Univ
HM 57				1600	Haryana

National average yield: 102 kg/ha; Seed rate:10–12 kg/ha; spacing 30 cm × 15 cm. 12/15 kg spacing:30 cm × 69 cm; Manuring and fertilizer application: Rainfed crop: 5–19 tonnes FYM/ha plus 30 kgP-/ha Irrigated crop: 20–25 tones FYM/ha plus 20–30–40 kg/ha of N,P and K

Propagation

Methi is propagated by seeds. and grown singly or mixed with other crops and grown. For seed and fodder, *desi (*native) methi is sown from mid September to mid November, and, for green leaves the sowing period can be extended up to March. *Kasuri methi* is sown during the same period particularly in December. In the hills, both types are sown during March–April and again in October. On the black cotton soils of Andhra Pradesh, sowing is done from mid-June to early July; and it is possible to grow a second crop in the same year during late monsoon. On these soils it is grown either pure or mixed with grams, gingelly or coriander.

For fodder crop the seeds are broadcast and then the surface is raked up to cover them well. If seeds are to be taken, the crop is sown in lines 20–30 cm apart. The seeds are sown through 4-tyned seed drills, the tynes being 24 cm apart, and the rows are covered by working the light-bladed harrow.

Seed Rate

It varies from 30 to 35 kg/ha. For Kasuri methi, it is 23–17 kg/ha and for irrigated crop it is higher.

Irrigation and Interculture

The plot is watered immediately after sowing, followed by another irrigation after a week. Subsequent irrigations should be after about 8–10 days. Weeds will not be there if the crop is planted thick, and, if any, are pulled out by hand. *Desi methi* germinates earlier than *kasuri methi.* Combination of eight irrigations with 15 kg N + 22.5 kg P205 (2 and 5 subscript please)/ha resulted in the highest yield of 7.17 q/ha in a fertilizer response investigation.

Tissue Culture

Multani (1981) investigated both diploid and auto-tetraploid strains of fenugreek which could regenerate complete plantlets from meristematic cells. Priyadarshan et al. (1989) have systematically reviewed tissue culture work on fenugreek and other important seed and tree spices.

Plant Protection

Control of Diseases Fenugreek is not affected severely by diseases. However, some of the diseases recorded are as follows.

Leaf Spot Caused by *Cercospora traverisiana* Sacc. which produces brown elliptic, or, round spots, later dark, olive colored growth in the center of the spot emerge.

Downey Mildew Caused by *Peronospora trigonellae* Gaum, which form greyish-white, powdery patches on the under surface of the leaves; the corresponding portions of the upper surface become yellowish or necrotic.

Powdery Mildew Caused by *Erysiphe polygoni* DC, *Oidium* sp., *Oidiopsis taurica* (Lev) Salmon. Dusting 300 mesh sulfur @25 kg/ha or spraying wettable sulfur or dinocap (0.25%) thrice at fortnightly interval, starting 40 days after germination. Controls the disease.

Rust caused by *Uromyces anthyllidis* (Grey) Schroet, producing small, round or oblong, dark brown pustules formed on both the surfaces of the leaves and on the petioles.

Damping Off of the young seedlings when the soil is too moist or there is excess rain. At Coimbatore, Tamil Nadu, India, drenching twice with Bavistin (0.1%) has been found to be the best treatment to control the root rot disease of fenugreek.

Some of the remedial measures which can be practiced are growing resistant varieties, spraying Bordeaux mixture and crop rotation.

Insect Pests The mite, *Tetranychus cucurbitae* (Rohman and Sapra) attack leaves which lose their green color, dry up and wither. The plants bear poor or no fruit at all. This is a serious pest in Punjab State of India. Aphids, *Aphis craccivora* Koch. and Myzus (*Nectarosiphon persicae*), are common insect pests throughout in Rajasthan state of India.

Harvesting *Desi methi* for vegetable purposes is ready for harvest 20 days after sowing. Plants are cut 2–3 cm above ground, kept cool and moist. The stubs produce new stalks ready for subsequent cut 12–15 days later. If harvested late, leaves taste bitter.

Seed Production *Methi* produces seeds freely. The cultural practices for vegetable crop may also be followed for seed production and details are given by Sharma and Arora (1991). Seed crop should be examined for weeds particularly *Melilotus alba* (*Senji*) and *Medicago hispida (Maina)*. The seed crop of *desi methi* is ready for harvest in 155 days. Plants are pulled out with roots and dried in the sun. Pods split easily, which are threshed, winnowed, further cleaned and dried in the sun for later storage.

Yield Yield depends on several factors, such as, whether the crop is rainfed or irrigated, grown for seed or herbage, the number of cuttings before seed -setting etc. A good main crop, irrigated, may yield between 200 and 250 q/ha of fodder. Seed yield can vary considerably. When grown entirely for seed, an average crop of common methi yields 15–19 q/ha of seed/ha and a crop of *kasuri methi* selection yields about 7.5 q/ha. But, when the crop is grown for leaf as well as seeds, the seed yield is about 2.5 q/ha for *desi methi* and about 1.5 q/ha for kasuri selection.

Post-harvest Technology

Packing and Storage: Fresh methi, used as a vegetable, does not store well and should be used immediately, soon after harvest. The well-dried, however, can be stored for up to 1 year. *Kasuri methi* is available as dried leaves in packets, for use, throughout the year. The seeds can be stored up to 2 years (Table 2).

Composition of Leaves Fenugreek leaves and stems are also rich in calcium, iron, carotene (Vit A), and ascorbic acid. Although fresh leaves contain only 3.5% protein, but, on a dry weight basis, they are comparable to pulses.

Table 2 The seed composition of methi

Property	Content
Moisture	6.3%
Protein	9.5%
Fat	10%
Crude fiber	18.5%
Carbohydrates	42.3%
Ash	13.4%
Calcium	1.3%
Phosphorus	0.48%
Iron	0.011%
Sodium	0.09%
Potassium	1.7%
Vitamins	
Vitamin A	1401 IU/100 g
Vitamin B1	0.41 mg/100 g
Vt B2 (2 subscript please)	0.36 mg/100 g
Vitamin C	12.0 mg/100 g
Niacin	6 mg/100 g
Calorific value (food energy)	370calories/100 g

According to another source, fenugreek seeds also contain gums(23.6%), mucilage (28%), trigonelline (0.13–0.28%) and total saponin (1.7%)

Marketing

Assembling and marketing: There are some important internal markets in India for fenugreek products. The produce is carefully packed in gunny bags of 50–100 kg capacity, before delivery for marketing.

Grades and Specifications

The Agmark (Indian) and ISI (International) grades of seeds are designated as:

1. Special
2. Good
3. Fair

Processed Products

Fixed Oil Fenugreek seed contains about 7% fixed oil. The fatty acids consist largely of linoleic, linolenic and oleic acids. It has marked drying properties. The dried oil is golden yellow in color and is insoluble in ether. The oil has a disagreeable odor and bitter in taste.

Volatile Oil Fenugreek contains only less than 0.02% volatile oil, brown in color and slightly odorous. Its specific gravity is stated to be 0.871 at 15.5 ° centigrade. Only very little is known about its chemical constituents and the product has no commercial value in India.

Effect of roasting and cooking on nutritive value of fenugreek seeds:

The loss of ascorbic acid is the least in pressure cooking and the most in in stir-frying of vegetable fenugreek. The loss of ascorbic acid is more at room temperature than in refrigerator. Raw dry fenugreek seeds contain about 150 mg trigonelline and practically no nicotinic acid. If the seeds are roasted sufficiently to brown them, slightly, about two-thirds of the trigonelline is converted to niacin or nicotinic acid (vitamin). An increase in free and total niacin has been noted when fenugreek seeds are roasted in a shallow pan over open fire. Seeds are used in the preparation of typical Indian and Egyptian foods. Some foods are prepared with fermented dough. In all cases, there is an increase in total and free niacin after cooking. The addition of 5% fenugreek to corn results in a 30–35% increase in free niacin in the cooked products. Fermentation results in a significant increase in both free and total niacin.

The CFTRI scientists roasted fenugreek seeds at 150, 175 and 200 ° centigrades and found that light roasted seeds (at 150 °) were superior to medium (175 °) and dark roasted (200 °) with regard to flavor and and nutritive value. No appreciable loss of total N and crude protein occurred, but, there was considerable decrease in total and free sugars as roasting temperature increased. Of the five reducing sugars, three non-reducing sugars, and three reducing sugars were completely lost. However, roasting of seeds significantly enhanced flavor.

Oleoresin

Fenugreek oleoresin has become one of the most important export commodities globally.

Methi Leaves The average vit C content of raw and fresh methi leaves is 43.1%mg/100 g. After boiling in water or steaming and then frying, the vegetable loses 10.8% and 7.4% of the vitamin, respectively. Results of organoleptic evaluation shows that the most acceptable method of leaf preparation is steaming.

Uses

As Food or Flavorant: Fenugreek has been used both as a food and as a food additive. In Egypt and Ethiopia fenugreek is a popular ingredient of bread, known to the Arabs as "*hulba*" and as "*Abish*"in Ethiopia. In Greece, the seeds either boiled or raw are eaten with honey. In the USA the seed is used in the manufacture of chutneys and in various spice blends, but, its most important use is as fenugreek extract, the principal flavoring ingredient in imitation maple syrup. It is used in recipes like Hearty Vegetable Soup and fenugreek Beef Stew.

Fenugreek seed is mainly of interest as one of the principal odorous constituents of curry powder, and is specially used in making mango pickle in North and North West India, in particular.

In Medicine

Medical "*Papyri*" from ancient Egyptian tombs reveal that it was used (seeds soaked in water until swollen into a thick paste, which is then fed to the patients) both to reduce fever and also as food. Fenugreek is believed to help digestion and metabolism in general. Seeds are used in colic flatulence, dysentery, diarrhoea, dyspepsia with loss of appetite, chronic cough, dropsy, enlargement of liver, spleen, rickets, gout and diabetes. Seeds have also carminative and aphrodisiac properties. Systematic investigations at the National Institute of Nutrition (NIN) in Hyderabad, India, have confirmed the excellent usefulness of seeds in diabetic control.

As cattle feed and in veterinary medicine: Ground, fine and mixed with cotton seed, it is fed to cows to increase flow of milk. Sour hay is made palatable when mixed with fenugreek herb.

In cosmetics, dye, hair tonic and as cure for baldness: In the middle east fenugreek is recommended as a cure for baldness. In Java, it is used in cosmetics preparation. As paste of fenugreek seed, women in South India use it each week to apply to hair to give it a good shine and texture.

Leaves as a Vegetable Curry

In Indian diet fenugreek leaves are used in making curry. Also cooked with rice. Eating fenugreek leaves fried with ghee is reported to give relief from giddiness, stomach pain. Fenugreek leaf paste is used to reduce swelling in any body part.

Fixed Oil Uses The fenugreek seed contains a fixed oil which has an overpowering celery-like odor which is extremely tenacious, and in recent years has attracted the interest of the perfume trade.

Other Uses "Fenugreek paper paste" has been developed by the Cardamom Research Institute (Spices Board) to coat bamboo mats, for drying black pepper, and, for making better and more hygienic paper (Madhusoodanan et al. 1996).

References

Madhusoodanan KJ, Premkumar J, Hegde RV (1996) Fenugreek – paper Papte-coated bamboo mat – a simple way to clean and save pepper. Spice India 8(4):17–18
Mishra JP (1987) Harvesting economic yield of Methi Abstr No. 1–34. National Symposium on Spice Industries. AFST (1) Delhi Chapter (Ed. J. S. Pruthi). p. 29
Multani DS (1981) Plant cell culture in crop improvement. Proc Int Symposium (Calcutta), Dec. 1981. pp. 435–439
Priyadarshan PM, Radhakrishnan VV, Madhusoodanan KJ (1989) Tissue-culture of spice crops: Retrospects and prospects. Spice India 2(4):15–16
Sharma BK, Arora SK (1991) Fenugreek. Spice India 4(4):19–21

Mustard

Botanical name: (1)*Brassica nigra* Koch,(2) *Brassica alba* or *Brassica hirta* or *Sinapis alba* Linn.(3) *Brasscica juncea* (Linn) Czernjajev

Origin and Historical Background

Mustard belongs to the genus *Brassica* and family Cruciferae. *B.nigra* (Linn) Koch provides black or "Trieste"mustard seeds, *B.hirta* Moenchi or *B.alba* Boiss or *Sinapis alba* (Linn.) provides white or yellow colored seeds and *B.juncea* (Linn) Coss and Czern, "brown and oriental mustard seeds"(also known as "Indian mustard seeds").

Mustard is a globally used spice and Pythagoras, the famous Greek mathematician, had written about it five centuries before the birth of Christ, and, it is widely believed that it was used much earlier, centuries ago to this, in Africa and China. Jesus immortalized mustard when he spoke of the power of faith even if it were "not larger than a tiny mustard seed".

Early Romans liked to mix the sweet "must" of new wine with certain crushed seeds which they called "sinapis". The resulting paste was called" *mustum arden*" (hot must). As time went on, "*mustum ardern*" became "mustard" and the seed itself took the same name. Later, the Romans spread their fondness for mustard to Britain; preparations of mustard paste became more sophisticated. The town of Djon in France is said to have begun making its "prepared mustard", now internationally famous, as early as the thirteenth century.

Among the mustard's many claims to fame is the fact that it was the subject of what may have been the longest food advertisement in history - a 3000 Worder by Alexander Dumas, for the "mustard house" of Bornibus in Paris, in 1870. A colorful treatise on the history of "prepared mustard" is reproduced in the "Dictionary of Cuisine".

K. P. Nair, *Minor Spices and Condiments*,
https://doi.org/10.1007/978-3-030-82246-0_10

The modern era of mustard seed, however, began in 1720 when Mrs. Clements of Durham in England, found a way of milling the heart of the seed to a fine flour. This technique was later industrialized in England in the nineteenth century which became the standard method for processing the seed for use as a spice in cooking and also for the "prepared mustard".

Yellow or White Mustard

The seeds of *Brassica hirta* or *Sinapis alba* are slightly larger than the other types used in food, averaging about 3 mm. in diameter and flattened laterally. As the name above suggests, it is mostly of a pale straw color. This may also have a slightly pinkish cast and there can be an occasional light brown colored one. The major distinction of the yellow mustard seed is that its sharp taste stems from a non-volatile flavoring substance. This gives it a sharp tongue taste, but, not the aromatic pungency.

Yellow mustard is thought to have originated in Europe. To day, it is widely cultivated there, and the USA still imports it from countries such as Denmark and Britain. However, most of US supply now is from Canada and the Dakotas-Montana area.

Brown and Oriental Mustard Seed

Both types are from *Brassica juncea*. They share the same chemical make up, though, the average fixed oil tends to run somewhat higher in the oriental, while protein and fiber are typically higher in the brown. The seeds of both average about 2 mm in diameter. The seeds of brown mustard are reddish brown to dark brown in color. But, the oriental seeds are mostly yellow with occasional brownish ones. The important difference in the species (applying to both brown and oriental) is that the seeds contain a volatile oil which produces a very pungent aroma as well as bite. The effect is like that of horse radish, a sensation which shoots quickly in the sinuses. Chinese restaurant mustard, the hot English mustard and Dijon and German mustards, all exhibit this effect because all are made from brown or oriental seeds.

The *Brassica juncea* probably came from Africa, India and the interiors of China. Today, they are widely cultivated in Canada and the northern plains of the USA. In recent years, the oriental seed has been replacing the brown mustard in the Canadian and American growing areas, as the oriental seed is more economical.

Black Mustad Seeds

The *Brassica nigra* seeds are not unlike brown mustard seeds in outward appearance. They are about 2 mm or less in size but tend to be a little more oblong than spherical. This species is thought to be native to Europe and the Middle east and until comparatively recent times it was the main seed of commerce in Europe. Since World War II, however, browns and orientals have gradually replaced it because they can be grown more economically. The black mustard is of no consequence in the US market to day.

The Uniqueness of Yellow Mustard

No other spice (except garlic) is quite like mustard. Its flavor and pungency can only be fully experienced by triggering enzyme action which releases it. The most effective trigger is water at room temperature, although other low acid liquids like milk and beer also work. Acidic liquids such as wine, vinegar, and lemon juice are poor triggers (but good subsequent preservatives) of the flavor. Heat, as in cooking, can also impede the flavor release. As a result, it is best to moisten mustard powder with water and let it stand for 10 min before adding it to just any food preparation, including salad dressings. Incidentally, any powdered mustard will produce the fire of a Chinese restaurant mustard if it is wetted to a thin paste and let it to stand 10 min. However, it should then be used immediately, because, the pungency dissipates within a few hours unless vinegar or other acidic liquid is added.

Unlike other pungent spices, such as, black pepper and capsicums, mustard's bite does not build up or persist. Hot mustard consumed with an appetizer will not dull the appreciation of foods that follow. Conversely, low levels of mustard (in mayonnaise and sauces) actually stimulate the sense of taste.

Description and Distribution

Genus *Brassica* consists of over 150 species of annual or biennial herbs, several of which are cultivated as oilseed crops (rape, *sarson, toria* etc.,) or as a vegetable or fodder crop. The oil yielding Brassicas are predominantly cross-pollinated, constitute a group of about which considerable confusion exists regarding their identification and nomenclature.

The "mustard flour" of commerce is a mixture of two types of mustard seeds, brown or black mustard (*Brassica nigra*) and white mustard (*Sinapis alba*). Its condimental properties are largely due to the essential oil principles of these two seeds.

Area, Production and Export

No reliable data are available on area and production, but, there is an urgent need to improve production and productivity.

Crop Management

In India mustard is mostly grown mixed with wheat in north India and *ragi* (finger millet, *Eleusine coracana*) in the southern India, especially Karnataka State. A number of different species are grown in different parts of India of which some are grown exclusively for oil, others as vegetable, and some are grown for both purposes.

Soil Generally, loam and clay loam soils are suitable. For pure seed crop, the land must be prepared thoroughly.

Manuring About 20–30 tonnes of well rotted FYM/ha is applied.

Seed Rate and Sowing 5–7 kg/ha of seed is necessary. And they are sown broadcast in September–October, covered with a wooden plank and the spacing is 30 cm between rows and 10 cm between plants.

Irrigation Irrigation is applied immediately after sowing of seeds, and, subsequently at 10–12 days interval.

Interculture One or two weedings needed to smother weed growth.

Harvesting When the crop is grown for dual purpose, both for the seed and the foliage, the leaves are removed 25–30 days after sowing, and, care must be exercised to leave enough leaves for photosynthesis and development of the seed stalks. In about four and a half months the mustard seed crop is ready for harvest. Harvesting is done by pulling out the entire plant from the soil and when the pods lose their green shade they are dried on the threshing floor and the plants are beaten with wooden sticks to remove the seeds which are dried and stored. About 1000–2000 kg seed is obtained per ha. With high yielding varieties and better cultural practices higher seed yield can be harvested.

Plant Protection

Insect Pests Aphids start attacking the plants from January onwards, which can be controlled either by using a systemic insecticide, such as, Phorate, applied with irrigation water, which is an effective control. If the crop is for seed purposes periodic sprays of malathion or folidol can be done to control the aphids. If the crop is for vegetable leaves, less poisonous sprays of pyrethrum extract or nicotine sulfate can be done.

Post-harvest Technology

Packing and Storage Seeds are cleaned by winnowing, packed in gunny bags and stored in a cool, dry place.

Quality Evaluation

In white mustard the seeds also contain the glucoside, "sinalbin", and the enzyme "myrosin". In the presence of water, myrosin hydrolyses into sinalbin, producing acrinyl isothiocyanate, which is only slightly volatile, sinapine acid sulfate and dextrose. The yield of volatile oil is low, 0.16% (Table 1).

In black mustard the volatile oil (specific gravity: 1.015–1.025; refractive index: 1.5267–1.5291) is obtained in a yield of 0.7%–1.2% after hydrolysis of glucoside, sinigrin, by enzyme myrosin (Indian seeds reported to yield 0.68% volatile oil). The oil is optically inactive and consists almost entirely of allyl isothiocyanate (93–99%). Specifications for pharmaceutical oil are (BPC): specific gravity: 1.014–1.025; n 20: 1.525–1.530; and allyl isothiocyanate content: not less than 92%.

Table 1 The composition of mustard seed

Property	Content
White mustard (*Sinalpis alba*)	
Moisture	7.2%
Protein	27.6%
Crude fat	29.7%
N-free extract	20.8%
Fiber	10.3%
Ash	4.5%
Black mustard (*Brassica nigra*)	
Moisture	7.6%
N-substances	29.1%
N-free extract	19.2%
Ether extract	28.2%
Crude fiber	11.0%
Ash	5.0%

For the preparation of volatile oil, the fixed oil is first expressed from seeds, which are subsequently macerated with tepid water for several hours and steam distilled. The oil obtained is an extremely powerful irritant owing to its volatility and penetrating power and is responsible for the painful nature of alcohol, or in the form with 50 times its volume of alcohol, or in the form rubefacient. It is also used in cases of pleurisy and pneumonia.

Brown Mustard(*Brassica juncea*)

The oil content of the seed of the seeds is usually 30–38%. Certain forms cultivated in Uttar Pradesh (India) and locally known as *Lahi, Lahta,* have higher oil content (42–43%). Commercial *Rai* also contains some *Sarson*. Some samples contain seeds of var. *Crispifolia* Bailey, which resemble those of var. *Regosa* (Table 2).

The volatile oil has the following characteristics:

Specific gravity: 0.995
Refractive Index: 1.5185
Optical rotation: 0 o (small o please) 12'

It is reported to contain allyl isothiocyanate and related compounds, among which crotonyl isothiocyanate has been identified.

Quality Standards

Quality standards have been prescribed under PFA Act. However, quality grading of condiment mustard needs to be implemented more vigorously. Admixtures of seeds of different species from different areas have to be discouraged to ensure uniform quality product.

Table 2 The composition of Brown mustard

Property	Content
Moisture	6.2%
Fat	35.5%
Nitrogenous substances	24.6%
N-free extract	20.4%
Crude fiber	8.0%
Ash	5.3%

Processed Products

The major processed products from condiment mustards and their admixtures in controlled conditions are as under:

Mustard Flour or Powder It is used in the manufacture of mayonnaise. It can also be used as a condiment in pickles, meat preparations and salad dressings. CFTRI, Mysore has developed an improved process for the production of "Mustard Flour or Powder" from mustard seeds. This powder can be used as in the preparation of "mustard paste", "table mustard" and other mustard-based products.

Compounded Mustard or Mustard Compound It generally consists of the flour of a mixture of seeds of *Brassica nigra, Brassica juncea and Sinapis alba*, and, in addition, it also contains powdered turmeric, starch or wheat flour (not exceeding 10%), and other spices. The Compounded Mustard patented by CFTRI, Mysore meets the specifications of UK and Australia. *B.juncea* is responsible for the pungent component of the flavor, while *S.alba* contributes to other flavor characteristics and its presence ensures that adequate quantities thioglucosidase are available for flavor generation. Both species produce intensely burning taste sensations on the tongue. Since 1970 *B.juncea* has replaced *B.nigra*. as a mustard component.

Recent investigations have proved a strong and positive relationship between sulfur nutrition of mustard plant and pungency or allyl-isothiocyanate content of mustard seed. These two findings should be of considerable interest both to the mustard processing industry and mustard grower, who should concentrate on *B.juncea* for mustard compound industry and should also ensure that the soil is not deficient in sulfur for producing high-pungency mustard. Any deficiency of sulfur in soil could be made up by the soil application of sulfates before sowing the crop.

Mayonnaise

One of the most important applications of the emulsifying qualities of mustard is in mayonnaise. Mayonnaise consists essentially of a creamy emulsion of vegetable oil in vinegar. The emulsifying agents present are mustard and egg yolk. Other substances are added for flavoring purposes, but, they do not materially affect stability. The specific role of these emulsifying agents has been discussed in scientific literature which shows that mustard not only is useful for its condimental qualities in mayonnaise, but, in addition, exerts a valuable emulsifying effect, which helps in the promotion of stability in mayonnaise.

Dried/Dehydrated Mustard Greens (Leaves)

Mustard leaves are dried or dehydrated, packed and sold in the market. Some of the Japanese mustard plants have large leaves and are exclusively grown as vegetables, which are also at times, dried/dehydrated, packed and sold.

Mustard for Industrial Buyer

The spice mustard comes in three forms for the industrial buyer, as follows:

1. White Mustard Seeds: The whole seeds are, select, bold (big) seeds of yellow type. They are used primarily in pickles, relishes and condiments, where appearance of whole seeds is an attraction.
2. Powdered Mustard/Mustard Flour:

This is the most important processed form of mustard seed. At retail, it is labelled "powdered", whereas industrially, it is often termed "flour", which is one of its designations in the FDA standards. The husk (bran) of the seed is removed by milling and the heart ("mildings") of the seeds are finely ground. The heart meat can be milled to a uniformly fine powder which disperses and bends particularly well in cooked foods, sauces and "prepared mustards". "Mustard Flour" is usually a blend of oriental/brown and yellow seed. The miller thus produces a product which can be adjusted to meet the buyer's end-use requirements. Industrially, the buyer purchases on the basis of specifications for such criteria as flavor (amount), heat and pungency levels, fineness of grind, color and protein content. Other factors, such as desired shelf-life for the end product, may be determined as well.

3. Ground Mustard:

   It is sometimes called "mustard meal"; thus, this is the product of grinding the whole seed, husk, included. It may be a blend of yellow and brown/oriental seeds, but, is mostly prepared from yellow mustard alone. It's principal customer is the sausage industry. The husk contains a mucilaginous substance which acts as a good binding agent in sausage products. The ground mustard thus specially offers this feature in addition to flavor.

   The US Department of Agriculture (USDA) limits use of ground mustard in sausages simply because of mustard's high protein content and the fact that protein content in total serves as one basis for government control of quality of sausages. Protein contribution of seasoning is restricted to 0.35% and this limits the use of mustard in sausage to about 1%.

Mustad Cake

When fixed oil is wholly or partially removed from ground mustard, the resulting by-product is termed "mustard cake". Since the oil removed is a vegetable oil, which does not contribute to flavor, the other valuable properties remain the same in the cake.

Handling and Storage of "Mustard Flour" Since the flavor of mustard is so well entrapped by nature, the storage requirements are not quite as critical as with other species. Ambient temperature and avoidance of excess moisture will be sufficient to give "mustard flour" a long shelf life.

Uses

As a Food Flavorant Mustard seed is an important condiment used to flavor various food products. Whole mustard is used as a flavoring material after frying, in curries, pickles and sauces as a garnish in salads. Mustard flour is widely used as a condiment to impart flavor and piquancy to all kinds of meat. It is also employed in the manufacture of propriety mustard powders, such as, "Coleman mustard". Besides, it is used in "prepared mustard", a paste product containing mustard flour, vinegar, spices etc. Most importantly, it is used in internationally famous mayonnaise, which consists of mustard and egg as emulsifiers, added to a creamy emulsion of edible oil and vinegar, containing some other flavorants.

Preservative Properties Investigations have found mustard flour, of all the spices and condiments, as the most efficacious in yeast fermentation of wines, followed by cloves and cinnamon. Little or no preservative action could be detected with cardamoms, cumin, coriander, caraway, celery seed, pimento, nutmeg, ginger, thyme, bay leaves, marjoram, savory, rosemary, black or cayenne peppers. Comparing the stronger of the above preservatives with benzoic acid and sulfur di oxide, it was found that mustard was definitely stronger than either of the two permitted chemical preservatives. Ground cloves have a potency of the same order as that of the benzoic acid, and superior to that of the sulfur di oxide.

Cassia and Cinnamon

Cassia (*Jangli dalchini*) and true cinnamon (*dalchini* or *darchini)* are very poplar spices commonly used in the Indian dietary. The term *dalchini*, which has come to stay in the Indian parlance, is actually a derivative from the Arabian term "*Dar-al-chini",* meaning, the wood (or bark) of China- one of the oldest and largest producers of Cassia, so well known in global commerce. Several reports indicate that cinnamon is the oldest spice known to mankind. Records show that it was known to the Egyptians even 2000 years BC. European and Arabian travelers/writers confirm the trade in cinnamon and Cassia from Sri Lanka, Seychelles, China, Vietnam, Madagascar, Indonesia and West Coast of India to other countries. Chinese motioned them in their earliest herbals, as well.

Even to day cinnamon and cassia occupy an important position in the world trade. In pre-war years, trade in cassia used to be more important than in cinnamon, but, the situation has now changed. The demand for cinnamon and cassia has steeply increased, and the second most important importer of cassia is the USA. Cassia is generally cheaper and considered inferior to cinnamon, and, therefore, is used as a substitute. Cassia and cinnamon, when whole, can be distinguished from one another. But, when ground, they cannot be distinguished from one another, giving way to admixture or adulteration of cinnamon.

Nomenclature

Cinnamon and cassia of commerce consist of layers of dried inner barks of branches of a number of species of ever green tropical trees belonging to even the same species known by 6-10 different names in different producing countries. However, after careful prolonged discussions by the member countries under the aegis of the ISO, the following species have been officially recognized (Table 1).

K. P. Nair, *Minor Spices and Condiments*,
https://doi.org/10.1007/978-3-030-82246-0_11

Table 1 Nomenclature of Cassia and Cinnamon

Popular English Names	Botanical Names
A. Cassia	
1. Cassia China or Cassia	*Cinnamom aromaticum* C.G..Nees or *C. Cassia*
2. Indonesian or Batavia Cassia or Java Cassia or Cassia vera or Padang Cinnamon	*Cnnamomum burmanni* Blume
3. Saigon Cassia (from Korea)	*Cinnamomum laureirii* Nees
B. Cinnamomum or True Cinnamon	*Cinnamomum zeylanicum* Blume

Species of Cinnamon Grown in India

So far as India is concerned, the most important aromatic cinnamon and cassia barks which are in commercial use are:

1. The "Genuine" or "True cinnamon" (*C.zeylanicum*) cinnamon
2. Karuva or "*Jangli dalchini*" (*C.inners*) cassia species
3. *Tejpat* or "*Tamal patra*"or Indian *Cassia lignea* (*C.tamala*) cassia species
4. *Tezpat* or Ram *tezpat* (*C.obtusifolium*) cassia species

The international nomenclature, Indian names, description and uses of the above three cassia species are briefly described below. Their method of cultivation is similar to that described under cinnamon that follows.

Jangli Darchini (Cassia) or Kuruva
Botanical name: *Cinnamomum inners* Reinw.
Family: Lauraceae

Description and Distribution

Cassia cinnamon (Cinnamomum aromaticum Nees.Syn. C. cassia Bercht & Presl.)

(Lauraceae) is a closely related species *C.verum* Bercht & Presl., the true cinnamon. Cassia bark, bark oil, leaf oil and oleoresin are the products of commerce and are important in the international trade. Cassia is also obtained from various sources like *C.cassia* Bercht. & Presl. *C.bumannii* (C.G. and Th.Nees) Bl., *C.laureirii* Nees and *C.tamala* Nees. These dried inner barks have a more intense aroma with a high essential oil content (1–4.5% as against 0.5–3% in cinnamon bark). Unlike cinnamon, various parts of *C.cassia* contain only one type of essential oil in which cinnamaldehyde is the main component. Leaf oil content of cassia ranges from 0.7–2% (Krishnamoorthy and Rema 1994).

The large evergreen tropical tree is reported to grow in the evergreen forests on the Western Ghats from Mysore (Karnataka State in India) and Coorg (also in Karnataka State) to Annamalais (Tamil Nadu in India) and Kerala State, in India.

The method of its cultivation, preparation, and, curing is the same as in the case of cinnamon.

Crop management/Propagation

Cassia is mainly cultivated in China. Only very few trees of *Cinnamomum aromaticum* are available in India, and, the existing ones are being replaced by other crops at present. Non availability of good planting material is a limitation for its cultivation and experimentation. Hence, an attempt to multiply them was made at IISR.

Propagation of Cassia Cassia is propagated through seeds, vegetative propagation has not been reported.

Air Layering in Cassia Air layering in 2-year old cassia plants (*Cinnamomum aromaticum)* is reported for the first time (Table 2). A success of 87.5 and 50% rooting was obtained when carried out during July and November, respectively, with 100% field establishment (Krishnamoorthy ad Rema 1994).

This method has a great potential since air layers ensure genetic uniformity and early maturity of planting material.

Uses

The end use of the bark is the same as that of cinnamon. The bark contains 0.5% volatile oil with the odor of cloves. Its wood is light brown in color, fairly hard and shining. It machines well, takes on a good finish and hence can be used for the manufacture of light machine.

Tejpat or "Tamal Patra" or Indian *Cassia lignea*

Botanical name; *Cinnamomum tamala* Nees and Eberm

Table 2 Response of Cassia to Air Layering

Month	Rooting (%)	Field survival (%)	No of roots				Length of primary roots (cm)	
			Primary		Secondary			
			Mean	Range	Mean	Range	Mean	Range
July	87.5	100	6.4	5-7	12.1	12-13	6.2	3-8
November	50.0	100	4.0	2-8	6.2	1-15	4.8	5-9

Source: Krishnamoorthy and Rema (1994)

Family: Lauraceae

Description and Distribution

Tejpat is a moderate-sized evergreen tree, attaining a height up to 8 m and a girth of 1.4 m. It is distributed in the tropical Indian Himalayas 915–2440 m above mean sea level. msl, Khasi and Jaintia Hills (915–1220 m above msl) and in the Indian states of Bengal, Meghalaya, Assam, Manipur, Arunachal Pradesh, Tamil Nadu and Karnataka.

The tree is a source of tejpat leaves, which are extensively used in northern India as a flavorant spice. The bark of the tree, known as Indian cassia bark or Indian *Cassia Lignea*, is collected from the trees growing in the foothills of Sikkim, Himalayas. There are regular plantations of *C.tamala* in Khasi and Jaintia Hills, Garo Hills, Mikit Hills, Manipur and Arunachal Pradesh in India. Tejpat is grown mainly in the Jaintia *Paraganas* of Sylhet district. Many plantations of tejpat in this tract are self – sown, a few are planted.

Area, Production and export Although precise statistics are not available, the total area covered is approximately 240 ha, mostly in the north-eastern India, distributed largely in the Himalayas at a height of 250–700 above msl.

During 1990–91, tejpat worth over 3.8 million Indian Rupees, INR (75 INR = 1 US$, as per current international exchange rate), and during 1992–93 INR 3.14 million worth of tejpat was exported, compared to INR 11.34 million worth of cassia exported, all from India.

Crop Management

The method of cultivation is similar to that of cinnamon, described earlier.

Harvesting and Collection of Tamala Leaves

Leaves are ready for harvest when trees are 4–5 years old, and the trees continue to bear for a century. Leaves are collected every year from young vigorous plants and in alternate years, from old and weak ones. Collections are made in dry weather from October to November until March. Continuous rain diminishes the aroma of the leaves. After harvest, the leaves are dried under shade for 5-7 days, following which they are ready for packing and marketing.

Preparation for Marketing

Small branches with leaves are dried in the sun for 3–4 days and tied up into bundles for marketing. Often, the leaves are separated and packed in cylindrical bamboo nets called "*bora*" or "*gungra*".

There are, at present, no quality standards for tejpat. Recently quality standards have been proposed for three grades of tejpat based on their systematic physico-chemical analysis of 41 samples collected from local Nagpur (Maharashtra State) market and other north Indian markets. These standards await approval of the Government of India (Agmark) and its Gazette notification.

Yield A tejpat tree yields about 10–25 kg of dry leaves/year.

Essential oil from Tamala leaf: An essential oil (0.7%) having a refractive index of 1.4791 obtained by hydrodistillation of fresh leaves of *Cinnamon tamala*, cultivated in North Cacher hills of north east India, was examined by chromatography. Fourteen oil components were identified which constituted 93.44% of the oil. Linalool was the main component (60.73%), whereas eugenol and cinnamic aldehyde were detected in trace amounts (<1%). Other components of significant occurrence (3%) in the oil were alfa(please use symbol)-beta (please use symbol) pinene (10.54%), beta (please use symbol) limonene (3.21%),beta(please use symbol) -pinene (10.42%), limonene (3.21%) and camphene (3.06%).

Uses

Tejpat leaves are mainly used as a spice flavorant in the preparation of the famous Indian vegetarian rice based dishes *Pulao* and *Biryani* (rice-meat based). In Kashmir, the Indian state, they are used as a substitute for chewing *pan* (Indian betel and arecanut based). It takes in Indian cookery the place of bay leaves, which are used in Europe. Used also as a clarifier in dyeing with myrobalans.

Leaves of *C.tamala* are carminative, used in colic and diarrhoea. They are hypoglycaemic, stimulant, anti-rheumatic, and an antidote against scorpion sting (Hussain et al. 1992). The bark is aromatic, coarser than the bark of cinnamon (C.zeylanicum) and is one of the common adulterants of true cinnamon.

Type Vietnam Cassia It occurs as a greyish-brown bark in a single or double quills, varying in length from 150–300 mm, about 10–38 mm in diameter and up to 6 mm in thickness (Table 3).

Table 3 Classification of Type Vietnam Cassia

Commercial designation of the grades	Physical characteristics of the bark
Whole rolls of cassia	
1. Thin rolls	Rolls up to 1.5 mm in thickness. The thin bark is slightly rough, dark brown and shows longitudinal wavy ridges, scars and numerous wart-like protuberances, 1.5–3 mm in thickness
2. Medium rolls	1.5–3 mm in thickness
3. Thick rolls	3–6 mm in thickness. The thick bark is lighter and more gray in color, very rough and with no wavy ridges
4. Broken pieces of Cassia	Small pieces resulting from trimming, sorting, handling and packing of rolls

Physico-chemical Specifications of Cassia Types

Cassia whole, in quills, quilling as whole bark and in ground (powdered) form shall comply with the requirements shown in the following table (Table 4)

Packing

Whole cassia or cassia powder shall be packed in clean, sound and dry containers made of material which does not affect the product or its flavor and protects it against moisture and loss of volatile matter.

The different types of whole cassia are usually packed as follows:

1. Type China Cassia: Cube-shaped straw mat bales of about 50 kg capacity
2. Type Indonesia Cassia: Bales of 50–60 kg capacity
3. Type Vietnam Cassia: Bales of 30–60 kg capacity

Storage The containers of Cassia should be stored in covered premises, well protected from moisture and direct sunlight, especially excessive heat. The storage place should be dry, free from objectionable colors and well proofed against entry of insects and vermin. There should be good ventilation in dry conditions and the storage room should be closed during damp conditions. In a storage warehouse, suitable facilities should be available for fumigation.

Transportation The containers should be handled and transported in such a manner that they ae protected from sun, rain, and, other sources of excessive heat, from objectionable odors, and from cross infestation, especially in the holds of ships

Table 4 Physico-chemical requirements of different Cassia types (China, Indonesia and Vietnam)

Physico-chemical characteristics	Requirement for			Test method
	Type China	Type Indonesia	Type Vietnam	
1. Moisture content (%) (m/m), max for whole and ground Cassia	12.0	12.0	12.0	ISO939
2. Total ash(%) (m/m) on dry basis, max.	4.0	6.0	4.5	ISO928
3. Acid-insoluble ash %	0.8[2]	2.0[3]	2.0	ISO930
4. Volatile oils ml/100 g	1.7[2]	1.0[3]	3.0	
Method yet to be finalized on dry basis (min)				
(a) Whole cassia[1]				
(b) Ground cassia[1]				

ISO.6538-1982. International Organization for Standardization (ISO)

[1]Values are tentative, pending preparation of test method

[2]This value relates to Kwantung cassia. Kwagesi cassia has a minimum volatile content of 3.5ml/100g

[3]This value relates to Padang cassia. Korintje cassia has a minimum volatile oil content of 1.3ml/100g

References

Hussain A, Virmani OP, Popli SP, Misra LN, Gupta MM, Srivastava GN, Abraham Z, Singh AK (1992) Directory of Indian medicinal plants. Central Institute of Medicinal and Aromatic Plants (CIMAP), Lucknow, India

Krishnamoorthy B, Rema J (1994) Air layering in cassia. J Spices Aromatic Crops 3(1):48–49

Cinnamon or True Cinnamon

Botanical name: *Cinnamomum zeylanicum* Blume
Family: Lauraceae

Description and Distribution

Cinnamon, Nutmeg, Clove and Allspice are the important tree spices grown in certain pockets in southern India, namely, Kerala, Karnataka and Tamil Nadu. Some years ago, the production of tree spices in India was just about 5000 tonnes, while the requirement was around 8000 tonnes. India used to import about 2500 tonnes of clove, 300 tonnes of nutmeg, and 60 tonnes of cinnamon. This caused a huge drain on the national exchequer. Hence, there is a tremendous scope for raising tree spice crops in the western and eastern ghats of southern India, north east region and Andaman islands.

Cinnamon was one of the earliest spice crops enjoyed by human kind, primarily as a food flavorant, a medicine, and as a perfume. The Egyptians were importing cinnamon early 2000 years BC. Wealthy Romans enjoyed cinnamon-scented baths. Every medieval magician kept cinnamon on hand and it was one of the ingredients of "love potions". The Chinese mentioned it in their herbals, and, even now, use it as a medicine.

Cinnamon is one of the most important tree spices of India. Like its cousin cassia, cinnamon consists of layers of dried pieces of the inner bark of branches and young shoots from the evergreen tree *Cinnamomum zeylanicum* Blume, which are obtained when cork and cortical parenchyma are removed from the "whole bark". The thickness of the bark ranges from 0.2 mm to 1.0 mm. Pure cinnamon is free from any admixture with cassia, which is considered inferior to the former in appearance, flavor and odor. As stated earlier, cassia is the commonest substitute for

K. P. Nair, *Minor Spices and Condiments*,
https://doi.org/10.1007/978-3-030-82246-0_12

cinnamon. While it may be possible to morphologically distinguish one from the other, in the whole form, it would be impossible when in powdered form.

The quality of cinnamon depends, among other factors, on the region where it grows. Sri Lanka and Seychelles islands grow the best cinnamon in the world. In India, cinnamon is grown in the Western Ghats. At Anjarakandy, Kannur district, in Kerala State, India, there is a 248-acre Randa Tara cinnamon plantation – one of the biggest, perhaps the oldest, plantations in Asia, established by the British.

Area, Production and Export

Cinnamon is mostly grown in Kerala State, India, but, reliable statistics on area, production and export are unavailable for both cinnamon and cassia. The demand for both is much more than what the country produces. To give an example, as early as 1981–82, 942 tonnes of cassia and 254 tonnes of cinnamon valued at Rs (INR). 33.4 million (1 US $ = 75 INR) and Rs. 8.58 million, respectively, were imported into India. Besides, India also imported 14.56 tonnes of cinnamon leaf oil and 3.4 tonnes of cinnamon bark oil during the same year, at a value of Rs.640,000 and Rs.249,000 respectively according to the Directorate General Commercial Intelligence and Statistics, Kolkota, West Bengal State, 1986. However, during 1987–88, only 1525 kg of cinnamon was exported out of India, valued at Rs. 47,205 according to Spices Board, India. This amply proves the dismal import-export scenario of cinnamon, of India, which draws attention to the urgent popularization of the crop in India. The increase in acreage and production can occur in Assam State, North East Frontier Area (NEFA), and, Western Ghats, in Kerala State.

World Trade in Cinnamon

Largest supply of cinnamon in world trade is from Sri Lanka, the largest producer and exporter, followed by Seychelles, Madagascar, and, India. According to the data supplied by ITC (International Trade Commission) trade in cassia and cinnamon, combined, was 25,000 tonnes, valued at US $ 35,000, as early as 1982. Of this trade, USA singly, imported cassia and cinnamon worth about US $ 10500, followed by Singapore (US $4530), Mexico (US $ 4111), Japan (US $ 3634), Hong Kong (US $ 3335), and Germany (US $ 1930).

Global Market Outlook

The world demand for both cinnamon and cassia is growing. It would be very prudent for India to keep this in mind and produce more cinnamon and cassia, keeping an eye on the quality, suitable for export and domestic consumption.

Crop Management

Since both cinnamon and cassia have similar requirements of soil, agro climate, and, agricultural practices, the cultural practices described below are applicable to both the crops.

Soil and Climate: Both cinnamon and cassia flourish in humid tropical regions receiving an annual rainfall of 150–250 cm with an average ambient temperature of 27 degree centigrade. Prolonged dry spells without rains are unfavorable for their growth. They can come up in places from almost sea level to an elevation of 1000 m above msl.

Cinnamon is a hardy plant which can be grown in almost all types of soils under a wide range of climate. The quality of the bark is highly influenced by the soil and ecological factors. In Sri Lanka, the world's leading cinnamon producer, it is grown in white sandy soil. Well drained deep sandy soils rich in humus are ideal for cinnamon cultivation. The best economic produce is obtained when grown in siliceous sandy soils. Waterlogged and marshy soils must be avoided (Krishnamoorthy and Rema 1987).

Propagation The following methods are used for propagation:

1. Through seed, which is the easiest and most widely adopted method. It is usually necessary to bag the fruits which are either picked from the tree or the fallen ones are collected and used. Ripe purple colored seeds would be ready in June-July. Seeds are removed from mother trees with desirable characteristics, such as, smooth bark, erect stem, ease of peeling the bark, vigorous growth, free from pests and diseases and with other good qualities like sweetness, pungency and flavor (Bavappa and Rethinam 1981).
2. It can also be propagated by planting cuttings and layers.
3. In old / established plantations propagation is done by transplanting old cinnamon tree-stumps, which are dug out along with as much soil and roots as possible and planted in sites already made in the field. They are given good shading and daily watering for 2–3 weeks when buds sprout. Greater care is required when digging out tree-stumps to ensure that no injury is caused to the roots, which will cause the stumps to die. The main advantage in this method is that the twigs would be ready for harvest in 1 year as compared to 3 years with seedlings.

The normal sowing method details in seed bed are given in the article of Bavappa and Rethinam, cited above.

Transplanting of Cinnamon into Polythene Bags

The seedlings are transplanted into polythene bags (30 cm × 15 cm) containing a mixture of soil, sand and well decomposed farm yard manure, in the ratio of 3:3:1 respectively, when they are 15 cm tall. The seedlings are ready for transplanting in 1 year.

Propagation of Cinnamon by Cuttings

Vegetative cuttings of cinnamon with 1 or 2 leaves can be made to rot within 40 days under humid conditions in a closed bin propagator with diffused light. Seradix could hasten their rooting. Cutting kept in a moist coir dust in a polythene bag roots within 60 days (Krishnamoorthy ad Rema 1987).

Transplanting Seedlings

Seedlings are transplanted in the main field when they are about 10-12 months old at a spacing of 2 feet between plants and rows. The planting is done in pits of 60 × 60 × 60 cm, dug earlier, and filled with compost and top soil. Most appropriate time for planting is June-July, which takes advantage of the South West monsoon in India. Planting can also be done in October-November during the North East monsoon. In the initial period it is necessary to provide shade. Weeding and mulching reduces weed growth; about 3–4 weedings required in the first 2 years, and thereafter one weeding in a year. Weeding in June–July and October–November and digging in August-September is done in the Anjarakandy Randa Thara Cinnamon Estate in Kannur District, Kerala State, referred to, earlier in this review.

Manuring and Intercultivation

Since leaves and shoots are the economic parts of the cinnamon tree, liberal application of fertilizer is made to encourage vegetative growth. N-P-K in the ratio of 2:1:1 is recommended (Bavappa and Rethnam 1981).

Fertilizers to supply 20 g N, 18 g P and 25 g K/seedling is applied in the first year after transplantation and this is doubled in the second year. Every year the dose of fertilizers is increased until the plant is 10 years old, when a single plant will receive 200 g N, 180 g P, and 200 g K. Organic manure is applied in a single dose at the commencement of South West monsoon, in May-June, while chemical fertilizers are split into two doses in equal parts, first at the time of applying organic manure in May-June and the second in September-October.

Weeding

It is very essential to clear all weeds from the planted area. Three to four weedings are required during the first 2–3 years after transplantation of the seedlings.

An investigation carried out to find out the relationship between the colors of young flushes and quality parameters revealed that the plants with purple colored flushes yielded about 20% more bark oil than the ones with green colored flushes. Both flushes were at par with regard to bark oleoresin content and leaf oil content. This indicates that flush color can be considered as a criterion to select quality of seedlings at the nursery stage itself (Krishnamorthy et al 1988). Sweet taste with aromatic fragrance, pale brown and thin bark with high yield is the ideal cinnamon type.

Pruning/Cropping

During the initial stages after transplantation, growth may be slow, but, as the plants advance in age, it becomes faster. When plants are 2-years old, they are cut back in June-July to a height of about 12 cm from ground level. The stump is then covered by earthing up. This is repeated for every side shoot, developing from the main stem, during the succeeding season so that the plant assumes the shape of a low bush of about 2 m in height, and a bunch of canes suitable for peeling would crop up in about 4 years time. Regular peeling operations can be commenced in the case of seedling bushes from the fourth or fifth year onwards, depending upon the extent of development of peeler shoots.

Variability Studies

Seventy one cinnamon accessions (germplasm) were investigated for variability quality parameters which revealed high coefficient of variation for dry and fresh bark yields, bark oleoresin, leaf oil, bark oil, leaf-size index, and percentage recovery of bark. Association analysis revealed significant correlation of fresh weight of bark and leaf oil with dry bark yield. Bark oil was negatively associated with leaf oil. The data on ranges of variation (CV% Coefficient of Variation) are summarized in the following table (Table 1).

Details of Elite Cinnamon Lines are given in the following table (Table 2).

Table 1 Mean, range and coefficient of variation (%) for 9 characters in cinnamon

Line characters	Mean value	Range		CV (%)
		Minimum	Maximum	
Leaf length (cm)	13.08	8.75	20.69	17.83
Leaf breadth (cm)	5.06	3.31	8.30	18.74
Leaf area index (LAI)	0.67	0.29	1.71	35.13
Fresh bark weight(g)	07.41	30.00	840.00	67.19
Recovery of bark(%)	32.08	10.70	80.00	34.00
Bark oleoresin(%)	8.48	1.32	20.02	57.08
Bark oil (%)	1.81	0.51	3.85	36.28
Leaf oil (%)	1.97	0.72	4.80	48.46
Dry bark weight (g)	64.70	8.00	305.00	74.73

Source: Krishnamoorthy et al (1992)

Table 2 Elite cinnamon lines based on quality analysis

Cinnamon	Elite line (numbers)	Leaf oil	Bark oleoresin Bark oil(%)	Cinnamon
Types/lines		(%)	(%)	Types/lines
Indian cinnamon				
1.	189	1.60	15.81	2.57
2.	203	2.16	14.70	2.85
3.	310	2.35	11.89	1.67
4.	312	2.73	19.14	2.00
Sri Lankan cinnamon				
5.	5	2.62	6.03	3.85
6.	44	3.41	12.45	1.58
7.	53	2.54	9.88	2.56
8.	65	2.75	8.58	0.95

Plant Protection

Insect Pests The cinnamon seedlings are attacked by certain pests. The important insect pests affecting cinnamon tree are:

1. Pink shoot borers
2. Leaf – eating caterpillars
3. Leaf ants
4. Termites

Pink shoot borers: They can be controlled by smearing the surface with BHC 50% Wettable Powder (WP) or Carbaryl is also effective in controlling the insect pest.

Leaf eating caterpillars (*Metisor plana*): They can be controlled by dusting 19% BHC. On older seedlings spraying 0.05% quinalphos helps. Leaf miners often infect tender leaves of seedlings, which can be controlled by spraying 0.05% monocrotophos.

Red Ants (*Occuphylla*): They sometimes make their nests by bringing together leaves of cinnamon plants. These ants thus spoil leaves and become bothersome when cutting the shoots for harvest. Dusting BHC 10% will control the pest.

New Records of Insect and Mite Pests: Seven species of insects were found to attack cinnamon plants in addition to those mentioned earlier. These are the following:

Fascellina castanea (Moore). This looper which appears during October-November is a defoliator. It is about 2.5 cm and is blackish-brown with three pairs of tubercles on the abdomen.

Hyposidra infixaria Walker. This is a greenish colored looper with white markings and feeds on leaves, seen in abundance in January.

Cleora alienaria Walker, subsp. *rasanaria* Swinhoe. This sub species which is endemic to Andamans is a defoliator. The larva is green with a light yellow head having a faint greenish yellow lateral band confined to posterior end. The anterior and posterior ends have black markings on the dorsal surface.

Lopharcha sp. The small green larvae of this month, measuring 0.5–0.7 cm, bore into terminal shoots resulting in wilting of affected plant parts.

Coleoptera: Chrysomelidae.

Dercitina sp. Beetles are small and black with a brown thorax. The adults make holes in leaves. Mostly seen in December.

Melanagromyza sp. The maggots are leaf miners, making dark brown tunnels in young leaves, occurring in December.

Coptosoma variegata (H.S). The green adults and nymphs are found in large numbers in April-May. They suck sap from tender shoots and fruits.

Diseases: The fungus *Corticum javanicum* attacks cinnamon during monsoon. Pale pinkish white crusts on stems, destruction of cambium layer and causing death of the shoot or twig are the main symptoms. Can be controlled by spraying 1% Bordeaux mixture.

Gray Blight Disease. This is caused by *Pestalotiopsis palmarum* in several parts of Kerala State.

Harvesting

Cinnamon harvesting starts from the fourth or fifth year after planting, depending on the availability of peeler shoots. Stems are cut during rains to facilitate peeling. The best time for peeling is when new flushes and leaves are hardened after rains.

Regular peeling operations commence from the fourth or fifth year after planting. The appropriate time for cutting shoots for peeling is determined with reference to the circulation of sap between the wood and the corky layer. The sap flow generally occurs soon after the onset of the monsoon. A test cut is first made, with a sharp knife, and if the bark separates readily, cutting should commence immediately. Best quality is obtained from the bark of shoots in the center from the middle portion of the shoot. From the first cut, after 3–4 years, 56–67 kg of quills/ha can be obtained, and, in a mature plantation, it can go up to 180–200 kg quils.

Post Harvest technology:

Preparation and Cutting of Bark Spice for Marketing:

Preparation and cutting of bark requires sufficient expertise. The following are the various steps involved:

1. Cutting of branches of the right size and shape in the right season at the right stage of maturity
2. Scraping of outer rough corky layer
3. Peeling and skilfully removing the bark from the inner wood
4. Piping and preparation of quills, quillings, feathering and chips

Cutting and Peeling

Cutting is followed by scraping and peeling operations. The peeling is a specialized operation peculiar to this industry which requires some technical skill and considerable experience. It is done by using a specially made knife which has a small round end with a projection to scrap the outer skin of the peel. First make a longitudinal slit from one end to the other. Then by working the knife between the bark and the wood, the bark is ripped off quickly. If there is any difficulty in peeling, the shoots are rubbed between wooden blocks to loosen the bark. This operation renders the peeling easier. The shoots cut in the morning are peeled the same day, without laying over to the next.

Preparation of Quils

The bark, as it dries, contracts and assumes the shape of a pipe, otherwise known as a "quil". The smaller quils or pipes are inserted into larger ones to form compound quils. They are then arranged systematically, end to end, in length of about 90 cm.

Drying of Quils

The quils are rolled by hand when they are soft and fresh and after rolling, they are dried on mats in shade, for about a week, depending on the weather and the type of the bark. When drying is complete, the bark is collected and packed into bundles of different grades for marketing.

The barks which are broken into pieces and those taken from very tender shoos are unsuitable for the preparation of quils. These are sun dried and graded as "quillings" and "featherings" respectively. Yet another grade of bark is the cinnamon "chips" obtained by drying the unpeelable bark removed in small chips from over matured shoots.

Bark Anatomy

Bark anatomy of four species of Cinnamon – three cultivated (*C.verum, C.cassia,* and *C.camphora)* and the most wild cinnamon, *C.malabatrum* have distinct differences as depicted in the following table (Table 3). Differences in bark structure are useful in distinguishing the genuineness of the true cinnamon (*C.verum*) bark and also the taxonomy of the genus (Shylaja and Manilal 1992).

Table 3 Bark characteristics of *Cinnamomum* species

Cinnamomum species	1	2	3	4	5	6	7	8	9	10	11
	(all in mm)										
C.verum	0.95	0.64	0.30	0.92	0.03	0.29	+	+++	+	++	++
C.cassa	0.97	0.60	0.36	0.65	0.03	0.30	+	+++	+	++	++
C.camphora	0.59	0.38	0.28	0.92	0.03	0.62	++	+	–	+	–
C.malabatrum	0.73	0.38	0.36	0.46	0.06	0.12	+++	–	+++	+	+
1:	Bark thickness										
2:	Thickness from epidermis to pericycle										
3:	Phloem thickness										
4:	Phloem ray frequency										
5:	Phloem ray width										
6:	Distance between sclerenchyma islands										
7:	Presence of phloem fibers										
8:	Presence of oil globules										
9:	Presence of raphides										
10:	Presence of crystalline inclusions										
11:	Presence of tannin in phloem rays										

Source: Shylaja and Manilal (1992)
+ Very sparse, ++Frequent, +++Very frequent, -Absent

Utilization of Leaves and Twigs

The leaves and very tender portion of the shoots are collected and used for extraction of leaf oil on a commercial scale. These are dried in partial shade for a day or two and then put in big metallic boilers for steam distillation of leaf oil.

Yield: The yield from one hectare of cinnamon plantation after attaining the proper yielding stage is 200–300 kg of bark and about 35 kg of leaf oil in an year.

Economics of Cultivation

Though cinnamon plants will start to yield from the fourth or fifth year, economic yields can only be obtained from the tenth year onwards. During the intervening period, however, the plantation may maintain by itself or provide a marginal income to the grower. The estimated expenditure in bringing up one hectare of cinnamon plantation up to tenth year is approximately INR 20,000 (1 US $ = 75INR). The expenditure, thereafter, will be about INR 5000 per annum. On the basis of an average yield of 250 kg quils and 35 kg leaf oil, the estimated gross income will be approximately INR 10,000/ha/per annum.

Grading

Cinnamon is available in the market in different grade designations in different countries, but, more common are the following four grades accepted by ISO and ISI, where grading is based on length, breadth and thickness of quils.

Quils: They are the long compound rolls of bark up to 1 m in length, which constitute the best grade.

Quilings: During the course of grading and transportation, some breakage of quils takes place. Besides, the smaller pieces which are left after preparation of the quils also go into this second quality cinnamon, known as "quillings". They are, of course, genuine cinnamon which are mainly used to grind and distillation of the cinnamon bark oil.

Featherings: The grade consists of the inner bark of the twigs and twisted shoots which cannot give straight quils or quillings of normal length. Thus, they are also genuine and are used in the same way as quillings. However, they often contain small chips.

Chips: These include the bark pieces obtained from thick branches and stems, trimmings of the cut shoots before they are peeled, shavings of outer and inner barks, which cannot be separated or which are obtained from small twigs by beating or hammering and odd-pieces of outer bark. They invariably contain more or less

Table 4 Variation in the composition of different cinnamon species

Property	Content
Moisture (%)	5.40–11.4
Volatile oil (%)	0.3–2.8
Fixed oil (%)	0.3–1.9
Fiber	25.6–30.5
Carbohydrates	16.6–22.6
Protein (%)	3.0–4.5

Note: The various properties depend on the region where the plant is grown, climate and soil type

inferior bark and woody material. The admixture is labeled "chips" which constitute the most inferior grade of cinnamon. Detailed quality standards have been laid down. The composition of cinnamon is given in the following table (Table 4).

Different Types of Cinnamon:
Type Sri Lanka.

This is the dried bark of cultivated varieties of *Cinnamomum zeylanicum* Blume (Family: Lauraceae). Cinnamon, Type Sri Lanka is produced in four forms as follows:

A. Quils
B. Quillings
C. Featherings
D. Chips

The details are given in the following table (Table 5).

1. Foxing: The occurrence of reddish-brown patches on the surface of the quils, which may become dark brown with time. Foxing can be (a) superficial ("mal-korachedi") and (b) heavy ("korachedi"). This sub-division is based on the depth of the patches
2. Extent is determined by visual examination
3. Bale: A package of any one particular grade of quils wrapped with suitable material for export

Type Seychelles

This is the bark of branches of *Cinnamomum zeylanicum* Blume, cultivated in Seychelles. It is produced in the four forms as depicted in Table 6.

1. Rough cinnamon Bark consists of slightly curved, elongated, irregular, medium or small pieces of the whole unscraped bark

Table 5 Grades/classification of cinnamon quils, Type Sri Lanka

Commercial designation of the grades and qualities	Diameter of quils mm, maxm.	Number of whole whole quils (1,050 mm) per kg, min	Extent of foxing (%) maxm.	Minimum length of quils in a bale	Pieces of tubes and broken pieces of the same quality per bale % (mm), maxm.
Aba	6	45	Nil	200	1
Continental					
C00000 special	6	35	10		
C00000	10	31	10		
C0000	13	24	10	200	1
C000	16	23	15		
C00	17	20	20		
C0	19	18	25		
Mexican					
M00000special	16	232	50		
M00000	16	22	60	200	2
M0000	19	18	60		
Hamburg					
H1	23	11	25		
H2	25	9	40	150	3
H3	38	7	65		

Source: SO:6539.1983. International Organization for Standardization (ISO)

Table 6 Grades/classification of cinnamon, Type Seychelles and Type Madagascar

Commercial designation of the grade	Physical characteristics of the bark
1. Whole tubes (full tubes)	Tubes of length of 15 cm and bark thickness up to 1 mm
2. Pieces of scraped bark	Broken pieces, rough and grooved scraped bark of thickness up to 2 mm
3. Pieces of unscraped bark	Broken pieces, rough and grooved, of width up to about 3 cm and length up to 20 cm. The bark can be up to 5 mm thickness
4. Chips, flakes and unscraped bark	Small pieces of unscraped bark of cinnamon stems

Source ISO: 6539-1983. International Organization for Standardization (ISO)

2. Scraped Cinnamon Bark which is obtained from younger shoots from bushes of the same species; the shoots are scraped with a curved knife before the bark is detached from the wood
3. Quils and Quillings which are prepared from the young shoots of bushes, in a way similar to that used for cinnamon, type Sri Lanka

Type Madagascar

This is the bark of the trunks or branches of *Cinnamomum zeylanicum* Blume growing wild in Madagascar. It is produced either as simple hollow scraped or unscraped bark of rather coarse appearance, about 30 cm long, from smaller branches cut with a knife or more usually in the form of unscraped or scraped pieces of bark from the larger branches and trunks broken off with the flat side of a hatchet (Table 6).

Extraneous matter: The proportion of extraneous matter in whole cinnamon should not exceed 1% (m/m) when determined by the method described by ISO 927. Extraneous matter comprises leaves, stems, chaff and other vegetable matter. The only mineral matter permitted are sand, earth and dust. In the case of Cinnamon quils, type Sri Lanka, take about 110 quils per bale of Continental grade and 230 g of quils per bale of Mexican or Hamburg grades, break them up and inspect the filing. Unscraped inner bark, scrapings, foreign matter, bark of wild cinnamon and other genera should not be present.

Physico-chemical requirements: Whole cinnamon and cinnamon powder must comply with the requirements shown in the following table (Table 7).

Table 7 Physico-chemical requirements of cinnamon of different types and cinnamon powder

Physico-chemical characteristics	ISO Requirements Common, Type Type Seychelles Sri Lanka	Type Madagascar	Test method
Moisture content (%) (m/m), maxm	12.0	12.0	ISO939
Total ash (%0) (m/m), ondry basis (max)	5.0	7.0	ISO928
Acid-insoluble ash (%0 (m/m) on dry basis (maxm)	1.0	2.0	ISO930
Volatile oil ml/100 g on dry basis			
Whole cinnamon (min)	1.0[a]	0.4[a]	ISO6571
Ground(powdered) cinnamon(min)	0.7[a]	0.3[a]	

Source ISO: 6549-1983, International Organization for Standardization (ISO)
[a]Tentative values

Packaging of Different Commercial Classes of Cinnamon Bark and Powder:

Whole cinnamon:

The different commercial classes of whole cinnamon are usually packed as follows:

1. Cinnamon type Sri Lanka: packed in cylindrical bales of about 45 kg each
2. Cinnamon type Seychelles and type Madagascar packed in bales of about 50 kg each
3. Cinnamon type Seychelles quills and quillings are packed in wooden boxes of about 100–150 kg each

Cinnamon powder:

Cinnamon powder is packed in clean, sound and dry containers made of a material which does not affect the product adversely or its flavor. Additionally, the containers should protect the cinnamon powder against moisture ingress and loss of volatile matter.

Packing of Cinnamon and Cassia and their products:

Cinnamon quils are exported in bundles packed in25 kg bales covered with jute sacking and bound with metal strapping. Cinnamon destined for grinding is packed in pressed bales to reduce bulk.

Cassia is also packed in 50 kg bales but is not compressed. Minimum quantities required are normally 5-tonne lots.

Cinnamon bark oil is packed in 10 kg capacity screw-top aluminum containers, two containers to a case.

Cinnamon leaf oil and cassia leaf oil are packed in epoxy-resin lined 180–200 kg capacity steel drums.

For smaller quantities, the oils may be packed in high-density polyethylene (HDPE) containers, but, lengthy storage is not recommended. Minimum quantities purchased by importers vary from as little as 20 kg for expensive bark oil to 1-tonne lots for the leaf oil.

The cassia and cinnamon oleoresins are usually packed in polyvinyl chloride (PVC) plastic pails of 15, 20 or 25 kg capacity each. Minimum quantities are variable but could be as little as 50–100 kg, although larger quantities are preferred.

Storage of Cinnamon and Fumigation

Containers of cinnamon should be stored in covered dry premises, well-protected from the sun, rain, and excessive heat. The store room should be dry, free from objectionable colors and insect and vermin proof. There should be good ventilation and the room should be closed during cold weather and damp conditions. If it is a storage house, there should be facilities to fumigate.

Transport of Cinnamon

The containers should be handled and transported in such a manner they are protected from rain, sun and other sources of excessive heat, from objectionable odors, and from infestation, especially during shipping.

Processed Products

The commercial products of cinnamon and cassia trees are (1) whole and ground cinnamon/cassia, (2) essential oils and (3) oleoresin which are briefly described below:

Cinnamon Bark Oil

Cinnamon bark oil contains 0.5–1% volatile oil, which can vary from 0.95% to 3.35%. The essential oil, generally manufactured in USA and Europe is steam-distilled mainly from cinnamon chips and refuse left over after the preparation of quils for the spice trade. Bark oil has the following properties:

1. Specific gravity at 15.5 degree centigrade: 0.0150–1.030
2. Optical rotation at 20 degree centigrade: 0–8 o (small o superscript please)
3. Refractive index at 20 degree centigrade: 1.565–15.99.

The oil is light yellow in color when distilled and changes to red when stored. It contains 60–75% eugenol benzaldehyde. Synthetic cinnamaldehyde and cinnamon leaf oil are common adulterants in bark oil.

Cinnamon Leaf Oil

Cinnamon leaves and dry very tender twigs/stems are collected, dried in partial shade for a day or two to expel most of the moisture and transferred to big stills in which steam under 15 lb pressure is passed through from the bottom of the metallic stills/boilers. This is continued for 7–8 hrs and the distillate after cooling is collected. The oil being heavier than water can be easily separated by decantation and filtration. The oil being aggressive/corrosive in nature, is packed and stored in dark brown glass containers (Pruthi 1998).

Cinnamon leaf oil has the following properties:

1. Specific gravity at 15.5 degrees centigrade: 1.045–1.055
2. Optical rotation at 20 degrees centigrade: -1 – +3 o(small o superscript please)
3. Refractive index at 20 degree centigrade: 1.530–1.548

Oil yield varies from 1.6% to 3.4%, has a pungent odor, hot taste and contains 70–80% eugenol and traces of cinnamic aldehyde. The oil is yellowish colored with a slight camphoraceous odor resembling that of clove oil. The various specifications for cinnamon leaf oil are given in the following table (Table 8).

Seasonal Variation in Oil Recovery

The yield and quality of cinnamon leaf oil is dependent on a number of factors, like the geographical origin of the leaf material, leaf cropping, distillation practice followed, and season.

The oil yield from different plants during the different seasons investigated varied significantly. It is presumed that summer is the best period for maximum oil recovery from cinnamon leaves (Shylaja et al 1996).

Cinnamon leaf oil of good quality is being manufactured at Anjarakandy Plantations, in Kannur Distritct, Kerala State, India (please refer earlier discussion).

Table 8 Physico-chemical specifications/requirements for cinnamon leaf oil

Physico-chemical characteristics	Requirements
Relative density at 27/27 degree centigrade	1.030.0–1.056.0
Refractive Index at 27 degree centigrade	1.525.0–1536.0
Phenol (as Eugenol) per cent by mass.min.	80
Aldehyde (as cinnamic aldehyde) per cent by mass.max	4
Water soluble materials	Nil

It has many pharmaceutical, perfumery and other allied uses. Ramachandra Rao et al (1988) have reported on the major constituents of essential oils in cinnamon leaves, grown in Odisha State, in Bhubaneshwar district, in India, which showed a wide variation from 0% to 89%.

Physio-Chemical/Quality Specifications for Cinnamon Leaf Oil

Physical Specifications

The oil should be a bright yellow clear liquid free from sediment, suspended matter, separated water and adulterants, when tested, It should be soluble in two volumes, of 70% alcohol (ethanol), occasionally with opalescence, when tested.

Cinnamon and Cassia Oleoresins

Cinnamon and cassia oleoresins are extracted by using a suitable organic solvent from commercial species and cassia, notably from chips. Bark is distilled for the preparation of oleoresin and the recovery is about 10–12%. It is a dark brown liquid containing 50% of volatile oil and has to be diluted before using it to flavor food items. Cassia oleoresin is normally produced from Indonesian cassia which is much cheaper than the Chinese cassia. Oleoresins are mostly manufactured in Western Europe and the USA. India has started to manufacture oleoresin and is now exporting it to overseas.

Other Products

The other minor products from cinnamon and cassia are (1) root bark oil, (2) cinnamon seed oil and (3) cinnamon buds for flavoring and spicing goods like quills and quillings.

Packaging and Storage

Packing: The material (oil and oleoresin) should be supplied in air-tight and preferably amber-colored glass or aluminum containers, permitting a minimum of air space, or, as agreed upon by the customer and the supplier.

Storage: The material should be protected from light, and stored in a cool and dry place.

Uses

Cinnamon bark is one of the most popular spices used all over the world. It has a delicate fragrance and a warm agreeable taste. It is extensively used as a spice or condiment in the form of small pieces or as a powder. It is aromatic, astringent, stimulant and a carminative, also possesses the property of checking nausea and vomiting. Cinnamon is used to flavor confectionery, liquors, pharmaceuticals, soaps, and dental preparations. Powdered cinnamon is a constituent of chocolate preparations especially made in Spain. Cinnamon is also used in candy, incense making, and perfumes. Cinnamon has also been found to help diabetics control blood sugar.

Stem Bark Oil

Bark oil is principally used in confectionery manufacture for flavoring, liquors, pharmaceuticals, soap and dental preparations. It has high germicidal activity (R.W.Coefficient of 14.0), but, on account of its irritant properties, it is not used as such. It is also a fungicide. It has the cordial and carminative properties of cinnamon without its astringency and is employed as an adjuvant in stomachic and carminative medicines. As a powerful local stimulant, it is sometimes prescribed in gastrodynia, flatulent colic and gastric debility. The dosage is 1–3 minims.

Fungitoxic Properties of Cinnamon Bark Oil

During screening of vapors emitted by different parts of higher plants, the bark of *Cinnamomum zeylanicum* exhibited the strongest toxicity, inhibiting mycelial growth of two sets of fungi, namely, *Aspergillus flavus*, and *Aspergillus niger* completely. The volatile antifungal fraction was isolated in the form of an essential oil which was standardized by physico-chemical properties and gas layer chromatography (GLC). The Minimum Inhibitory Concentration (MIC) of the oil was found to be 400 ppm against both the test fungi. The oil showed fungi-static nature at its MIC, but, proved fungicidal at higher doses. The fungitoxic potency of the oil was found thermostable and possessing long shelf life and was sustainable to heavy inoculum density. The oil showed broad range of activity inhibiting all the 35 storage fungi tested at its MIC (400 ppm). Thus, the oil may constitute an ideal, indigenous and effective fumigant preservative for protection of rice in particular and other food commodities in general.

Cinnamon Leaf Oil

The yield of cinnamon leaf oil varies from 1.6 to 3.4%. Cinnamon leaf oil equals clove oil in eugenol content (70–95%), which makes it useful in perfumery, cosmetics and flavoring industries. In USA, the cheaper Seychelles oil is used as a source of eugenol in the synthesis of vanillin, while Sri Lankan Oil, which is considered superior, is employed in soap manufacture. It is also used to flavor sweets and confectionery and is a common adulterant for the bark oil and in that respect, leaf oil competes well with clove stem and leaf oil. Cinnamon leaf oil is used to flavor sweets and confectionery. It is used as an embrocation in rheumatism.

Root Bark Oil

The root bark yields 3% oil which differs from both stem-bark ad leaf oils. It is a colorless liquid with a camphoraceous odor. It has a specific gravity of 0.994, optical rotation (alfa), please use symbol) D:+50 C. It contains camphor, pinene, cineole, dipentene, phellandrene, eugenol, safrole, caryophyllene, borneol, and possibly, cinnamic and benzoic aldehydes. Camphor separates out on allowing the oil to stand. The oil, however, is not an article of commerce.

Seed Oil

The seed contains 33% fixed oil, formerly used for making candles. The oil also called "cinnamon suet" is obtained by boiling crushed ripe fruits suspended in water. The oleaginous matter rises to the surface and solidifies on cooling.

Cinnamon Buds

Cinnamon buds are as good for flavoring and spicing as the bark itself.

Cinnamon Wood

Cinnamon wood provides a soft timber for use in low-grade board wood. The timber is moderately soft, not very strong, seasons without difficulty, but, warps, splits, cracks, and is liable to train. It is fairly scented, straight-grained, medium and fairly even-textured.

In conclusion, it can safely be said, that every part of the Cinnamon plant has an economic value.

References

Bavappa KVA, Rethinum RA (1981) Cinnamon cultivation and processing. Tech. Bull. 5, UNDP/FAO Research Project on Minor Export Crops, Department of Minor Export Crops, Sri Lanka

Krishnamoorthy B, Rema J (1987) Nursery techniques in tree spices. Ind Cocoa, Arecanut & Spice J 11(3):83–84

Krishnamorthy B, Sasikumar B, Rema J, Gopalan A, Abraham J (1992) Variability and association studies in cinnamon (*Cinnamomum veruum*). J Spices and Aromatic Crops 1(2):148–150

Pruhi, J.S. 1998. Cinnamon in: spices and condiments. National Book

Ramachandra Rao Y, Paul SE, Dutta PK (1988) Major constituents of essential oils of *Cinnamomum zeylanicum* L. Indian perfumer 32(1):86–89

Shylaja M, Manilal KS (1992) Bark anatomy of four species of *Cinnamomum* (Lauraceae) from Kerala. J Spices and Aromatic Crops 1(1):87

Shylaja MR et al (1996) Seasonal variation in the recovery of cinnamon leaf oil. Spices India 1:11–12

Trust, New Delhi, India 5th Edn (n.d.) (Revised, updated and enlarged to 340 pp.) 86–90, and Cassia *Ibid.* pp. 69–80

Clove

Botanical Name: *Eugenia caryophyllus* (Spregal) Bullock et Harrison
Syn: *Syzygium aromaticum* (merill et Perry)
Family: Myrtaceae

Description, Distribution, and Economic Importance

The clove of commerce is the air-dried unopened flower bud obtained from a handsome, medium-sized, evergreen, straight-trunked tree that grows in Kerala and Tamil Nadu States in India. The tree is myrtle which grows to a height of 10.7–12 m. begins to flower in about seven years from planting and continues to produce for another 80 years or more!

Clove is one of the most important spices in the world, as judged from the world trade, being next only to Black pepper (*Piper nigrum*). According to the survey conducted by the International Trade Center (ITC), Geneva in 1991, the total world export of clove averaged about US $ 34.9 million in 1990. Tanzania supplies about 90% of clove for export.

The term "clove" is derived from the French word "clov" or "clou" and the English word "clove", both meaning nail from the likeness of the unopened flower bud of the clove tree to a broad-headed nail. The bud is composed of two parts, the stem and the bulbous head. For sale as "whole cloves", it is most desirable for the buds to be intact and attached; but, this is not as important when they are required for grinding.

Clove is one of the most ancient and valuable spices of the Orient known as far back as 100 BC. This spice was known to the Chinese as far back as 266 BC. Clove was imported into Europe in 125. Its source and place of origin were shrouded in mystery until the Portuguese discovered the Molucca and Island of Indonesia in the sixteenth century. The French succeeded in 1770 in introducing the clove tree to

K. P. Nair, *Minor Spices and Condiments*,
https://doi.org/10.1007/978-3-030-82246-0_13

Mauritius and Reunion from where it reached Zanzibar. Clove was established in Sri Lanka in 1796, before the arrival of the British there. For some time clove was a Portuguese and after a Dutch monopoly like the Black pepper.

In India, clove was introduced by the East India Company (British) in 1800. By far, the biggest clove producing country in the world is Zanzibar, followed by Pemba, Madagascar and Indonesia. Clove is also produced in Malaysia, Sri Lanka and Haiti, but, not on a commercial scale. The major importer countries of clove today are USA, India, West Germany (now united Germany) and France. India depends almost totally on foreign countries for its clove needs, like Zanzibar.

Area, Production and Export/Import

Globally, the clove coverage was 276,000 ha in 1982, producing about 63,730 tonnes. Of this, Indonesia produced 42,000 tonnes. In India, cultivation of clove is primarily confined to southern India. As per the statistics provided by the National Research Center for Spices (now Indian Institute of Spices Research in Calicut, Kerala State, India), in 1995, India produced about 4000 tonnes/annum from a total area of 1865 ha. Since then, the institute has been working on a dwarf clove, which is expected to bring about a breakthrough in production.

What Are the Constraints of Clove Production in India?

The following constraints can be spelled out:

1. The very long pre-bearing age (the clove tree grows more than 80 years)
2. Lack of proper scientific knowledge about the crop management, in particular, in remote areas
3. Lack of availability of good planting material in remote areas, in particular

The above constraints result in clove demand outstripping production targets in India.

Crop Management

Site for Clove cultivation:

In southern India, the region that is chosen for clove cultivation should meet the following criteria:

1. Atmospheric and soil conditions: High temperature (25–35 degree Celsius) with bright sunlight, good and well distributed rainfall (above 150 cm annual), high

relative humidity (above 70%), well drained fertile soil, with sufficient moisture, throughout the year, are the four essential criteria for good clove cultivation.

2. Altitude or Elevation: Elevation 900 m above mean sea level is not conducive for good clove growth, as the plant will not receive sufficient warmth for robust growth. Likewise, if places below 500 m below mean sea level are chosen, unless there is sufficient rainfall during warm months from January to May, it would result in hampering good growth. The clove tree will luxuriously grow on any mountainous tropical island or place, where, as the saying goes, "it can see the sea".
3. Soil: The soil fertility could be manipulated by the addition of fertilizers or organic manure, but, it should be well drained, and, the pH should be between 4 and 6.5. Clove plant comes up well on slopey lands with good drainage.
4. Irrigation facility: Watering the trees in summer months from January to May and a short period in September on the western side of Western Ghats, in Kerala State, India, and most of the months in the eastern side of Western Ghats, is essential to maintain wet conditions around the tree. Since the tree is a water-loving one, the wet condition should be ensured around the soil layer just below the ground level. Hence, while selecting the site for clove cultivation, it must be ensured that there is sufficient water for irrigation, as well.
5. Atmospheric humidity: Atmospheric humidity is another criterion to ensure good clove cultivation. The relative humidity should be well above 70%.
6. Protection from wind: Cloves require a location protected from winds. If the site is "open" wind breaks must be in place; Dadaps and Silver oak planted in between clove trees provide adequate shade and also serve as wind breaks. Shade is very essential especially in the early stages of plant growth

Clove is a strictly tropical plant and requires a warm humid climate. Although there has been a general belief that the clove requires proximity to the sea for proper growth, and yield, experience in India has shown that the trees do well even in the hinterland conditions, as well. In fact, clove trees growing in sub-mountainous region have also been found to perform better. Humid atmospheric conditions and well distributed annual rainfall of 150–250 cm are other ideal requirements of the cop.

Varieties

The clove tree appears to be of one uniform type, and, no distinct varieties have been recognized. Differences have, however, been observed in tree shape (tall or spreading) bearing habit and cropping season; variations also occur in color, yield, shape and dimensions of cloves. The Penang cloves (from Malaysia) are one of the best, which should be grown wherever possible.

Propagation

Clove is propagated through seed which is called "the mother of the clove". Usually, the seeds are available from June to October; the seeds lose their viability within a week after harvest and hence it is necessary to sow them immediately after harvest. Only fully developed and uniform seeds are selected for sowing. Seeds collected from high yielding trees can be sown with or without the fleshy outer cover. If seeds are to be transported over distances, they are best kept in baskets made of palmyrah leaves and not in polythene bags as this will adversely affect germination. Removal of outer cover is done by rubbing seeds with sand or ash after soaking them in water for 24 hours. This facilitates early and better germination. Cover the top of the soil with grass and water twice a day and provide shade with thatches (1.5 m high) to avoid scorching. Seeds are sown at a spacing of 2–3 cm and at a depth of 2 cm.

Factors Affecting Germination

Clove seeds germinate in about a fortnight and it may last up to 40 days. Kumar *et al.* (1992) have reported that germination may last up to 84 days. Temperature plays an important role in seed germination. Kumar *et al.* (1992) have also noted that size of mother clove seed (husk removed) gives better germination as shown below (Table 1).

Highest percentage germination (97.5%) and lesser days (24) for germination were noticed in hulled mother cloves treated with 0.1% Carbendazim. Details ae given below (Table 2, Chezhiyan et al. 1996).

There was an investigation on pre sowing of seeds on germination percentage, the results are summarized in the following table (Table 2).

Table 1 Effect of seed size on germination

Seed size	Germination (%)	Days for germination
Large	50	70–84
Medium	40	71–84
Small	20	73–81

Table 2 Effect of pre-germination treatment on Clove seeds

Treatments	Germination (%)	Number of days taken for germination
Hulled seeds	80.0	26
Hulled seeds +0.1%		
Carbendazim	97.5	24
Unhulled seeds (control)	58	60

Source: Chezhiyan et al. (1996)

Raising Nursery

Raised nursery bed on fertile soil with a high percentage of organic matter is prepared in a shady place. Beds are 1 m wide and 2–3 m long. Seed should be placed 2.5 cm deep at a spacing of 12–15 cm. Seeds completely germinate in about 45 days.

Transfer of Seedlings to Bamboo Baskets/Mud Pots/ Polythene Bags

Six month old seedlings are transferred to any of the above and nurtured for 2 years. As the clove seedling roots are delicate, care must be bestowed while transplanting, preferably, on a rainy/cloudy day.

Transplanting in the Main Field

Transplanting is best done in June at the commencement of the south west monsoon. Pits measuring 60–90 cubic meters are dug at a spacing of 6 m both ways-plant density at 80/ha- about a month in advance of planting. Pits are filled with farm and/organic manure and closed with top soil. Watering must be done regularly for the first 2 years since planting. The seedlings must be provided with shade-using palmayrah/coconut leaves.

Propagation by Grafting

Clove can also be vegetatively propagated by grafting on its own root stock. But this type of propagation is not at all popular in India.

Mixed Cropping

Clove can be successfully intercropped with coconut/arecanut and the leaves of their trees will provide enough shade to the clove plant. Intercropping with pepper can also be taken up.

Weeding

Working on soil around the transplanted seedlings will keep the field free of weeds.

Support

It is important to prevent the stem or roots from breaking when the tender plants grow, hence, they could be tied with jute rope. This will also help the plants grow well against heavy winds, without breaking, which must be ensured for nearly 25 years when the plants attain the height of a tree.

Manuring

Initially about 15 kg organic/cattle manure may be applied, but, this should be increased to 40–50 kg/tree when the plants have grown for about 15 years. Additionally, 20 g N + 18 g P + 50 kg K/ tree is applied in the first year which should be increased to 40 g N + 36gP + 100 g K in the second year, which should be gradually increased to 300 g N + 250 g P + 750 g K.

Interculture/Thinning

This should be done to control weed growth and thinning will ensure that the plants are not too crowded with overgrown shoots.

Irrigation: Up to 2 years regular irrigation is required. Although the grown up trees can survive without water for long, it is advisable to irrigate on non rainy days. Each plant requires not less than 30 l water/day.

Soil and Moisture Conservation Work

If the clove garden has been raised on slopes, terraces are to be formed along the contour with 1.5 m wide inward slope. These terraces should be reformed once in 13 years. Further, every alternate terrace, or, preferably all terraces, are to be strengthened by contour wall. This will prevent loss of top soil.

Water Management

Since the plants are found naturally in the equatorial atmosphere, they require very good sunlight and copious watering to always keep soil moisture around the plant roots sufficiently high. Hence, when a clove garden is raised below 500 m mean sea level provide adequate irrigation facilities to ensure that the clove plant is not starved of water during summer months.

It would be advisable to construct/install a fairly big-sized water tank at ground level on the highest point in the field, pump the water into the storage tank from where it can be distributed through pipeline to all the trees. Both drip irrigation and sprinkler irrigation are not advisable for clove gardens unlike in coffee/tea gardens. It might appear that initial investment in these activities is high, but, it must be acknowledged that this is the only tree/plant from which economic revenue could be harnessed for almost 250 years. Additionally, it is possible to interplant Black pepper, which would bring added revenue, without hindrance to the growth of the clove plant until the clove plant enters its 20th or 25th year.

Harvesting and Curing

Clove plant begins to yield from the 7 h or eighth year after planting and full bearing age is attained after about 15–20 years since planting. It continues for 80 or more years. Flowering season is from September to October in plains and December to February at high altitudes. Flower buds are produced on young flushes. It takes about 4–6 months for buds to become ready for harvest. The optimum stage for picking clove buds is indicated by the change in color from green to slightly pinkish tinge or just before opening of bud. If they are allowed to flower, they have no economic value as a spice. The mature flower buds are carefully picked with hand. Care should be taken to pick buds at the correct time as otherwise the quality of the cured produce will be lost to a considerable extent. When the trees are tall and the clove branches are beyond reach, platform ladders are used for harvesting. Bending the branches or knocking down the bud-clusters with sticks is undesirable as these practices adversely affect future bearing of the tree as well as the quality of the buds.

In most cases, there are two crops in one year, the first one in late summer or early fall and the second in late fall of the first year. It is the nature of the clove tree to vary its yield from "bumper crops" to "light ones" in cycles, which range from 2–6 years. In volume of production, a mature healthy tree may produce from 3 to 18 kg in one harvest. It is no wonder that a 350-0ld year tree in Ternate Island of Molucca in Indonesia still yields about 75–250 kg of dry cloves per annum (Palani 1988).

Plant Protection

Insect Pests

Cricket: They sometimes cut tender seedings at the ground level and can be controlled by drenching with 0.05% BHC.

Scales: They infest tender shoots and leaves of older seedlings in the nursery, which can be controlled by spraying 0.05% monocrotophos.

Stem borer: This is the most important pest of clove, which bore into the stem resulting in the part of the plant where the attack happens to dry up and can be controlled by spraying 0.2% BHC.

Scales and mealy bugs: These suck the sap from the tender portions of the plant, resulting in the tender part drying up, which can be controlled by spraying 0.05% dimethoate or 0.05% monocrotophos.

Grasshoppers: Nursery seedlings are damaged by them, which can be controlled by spraying seedlings with 0.2% BHC.

Grub: Attacks the plant roots, and, the grubs originate from the compost or cattle manure applied. Application of thimet 10 g destroys the grubs.

Animals

Porcupine and Boar: Clove garden next to forests are targeted by wild boar and porcupine, which damage the bark completely around the collar portion of the plant, and the plant dies. Even 10–20 years old plants are not spared. Erecting chicken mesh tree guard 2–3 cm radius around the plants helps protect the clove plants from these animal attacks.

Plant Diseases

Die Back: This affects young seedlings and grown up trees alike. Leaves rot and fall, often leading to total defoliation of the plant. Spraying 1% Bordeaux mixture helps to control the disease.

Root Rot: This is caused by nematodes. Younger leaves from the top turn pinkish and leaves start to wither and fall off from top to bottom. Application of thimet or phorate immediately after observing the symptoms will help control the nematodes.

Collar rot: Seedlings in the nursery succumb to this disease. Spraying dithane M-45 controls the disease.

Nursery Pests and Diseases

Seedling wilt: Leaves of the affected plant lose their natural luster, droop, finally leading to the death of 5–40% of the seedlings. The fungi *Cylindrocladium* sp., *Fusarium* sp., and *Colletotrichum* sp. cause this disease. Spraying the foliage with Bavistin at 2 g/liter water helps control the disease.

Leaf rot: This is caused by *Cylindrocladium quinquiseptatum*. Symptoms start as dark patches from the edges of the leaves and spreads over the whole leaf leading to severe defoliation. Spraying Bavistin at 2/liter water helps control the disease.

Yield

The following factors determine the yield of the clove plant:

1. Age of the plant
2. Number of flower buds/inflorescence
3. Number of inflorescence branches
4. Size of the flower bud

A well maintained fully grown tree under favorable conditions of maintenance may give 4–8 kg of dried buds or more. The average annual yield after 15th year of planting may be taken as 2 kg/tree or 500 kg/ha in India.

Economics of Cultivation

Although the clove tree starts to yield from the seventh or eighth year after planting, economical yield is obtained only from the 15th year onwards. During the intervening period, the plantation may maintain by itself or provide marginal income to the farmer. The estimated expenditure to maintain 1 hectare of clove plantation to the bearing stage (eighth year) is INR 25,000 (75 INR = 1 US$) The annual maintenance charges, after the plantation has reached the full bearing stage (15th year) will be about INR 3500/ha. Considering that the average yield after this period is about 500 kg/ha, the gross income would have been INR 50,000/ha at the prevalent selling price in 1997. This would be much higher at current prices.

Constraints in Clove Cultivation

The following are the important constraints in clove cultivation:

1. Very low germination percentage

2. Slow germination of seeds
3. Slow seedling growth
4. Difficulty in harvesting the flower buds due to inaccessible height leading to 30–35% crop loss

The above constrains, in particular, the fourth one urgently demands that clove breeders develop short statured or dwarf plants.

Post Harvest Technology

Preparation of Clove for the Market

After removing flower stalks, the buds are spread evenly to dry in the sun either on grass mats or on cement drying floors. During nights, the buds should be stored under cover, lest they reabsorb moisture. Drying period depends on prevalent weather conditions. Normally clove buds dry well in 4–5 days, and, dry faster (in about 4 hrs), when dried under heated conditions in zinc trays over regulated fire. Fully dried buds are tan colored and their stems get a dark brown color and both are crisp. Improperly dried and stored cloves have much darker color and somewhat wrinkled appearance. This is inferior in quality and fetch low price. If the product is uniformly good, 8000–10,000 cloves would weigh a kilogram and 8–10 cloves would only weigh one gram. They are very light weight.

The fruit is about 2.5 cm long and 1.25 cm wide. The seed is oblong, rather soft in texture, and grooved on one side. The leaves, unripe fruits and broken cloves including the stalks are all aromatic and yield essential oil.

Grading

The air-dried cloves are graded according to their appearance and level of impurity present. Good quality cloves should be brownish-black in color with full and plump crown, somewhat rough to touch, without wrinkles and should not contain more than 12% moisture and 2–3% foreign matter (including *"Khoker", "Headless" and "Mother Cloves", "Clove Stems*").Yellowish wrinkled cloves are immature and yield volatile oil poorer in eugenol content – the main constituent. Cloves should have fine aroma and flavor, be free from mold and should readily exude oil when the stem is pressed with the thumb and finger nail – a layman's test before purchase of this valuable spice that one is not cheated with "exhausted" cloves, which are, cloves from which essential has been removed partially or wholly. Such exhausted cloves do not exude oil when pressed with the nail of the thumb and fore finger.

Grading of Whole Cloves

The details of grades of whole cloves are given in the following Table (Table 3).

Physico-Chemical Quality Requirements/Specifications

The details of the physico-chemical quality requirements and specifications are given in the following table (Table 4).

Ground (Powdered) Cloves

The physico-chemical quality requirements/specifications for powdered clove are given in the following table (Table 5).

Packing of Whole and Ground Cloves

Whole or ground cloves (powdered) should be packed in clean containers made of a natural material which does not affect the product and protect it from moisture ingress and from loss of volatile matter during storage/transhipment.

Table 3 Classification of whole cloves

Grades	Headless cloves % (maxm)	Tendrils, mother cloves % (maxm)	Khoker cloves % (maxm)	Extraneous matter % (maxm)
Special(hand-picked)	2	0.5	0.5	0.5
Grade 2	5	4.0	3.0	1.0
Grade 3	*	6.0	5.0	1.0

Source: ISO:2254–2960 International Organization for Standardization
[a]Not specified

Table 4 Physico-chemical quality requirements/specifications of whole cloves

Physico-chemical characteristics	Requirements	Test method
Moisture % (m/m), maxm	12	ISO 939
Volatile oil ml/100 g on dry basis (min)	Method under consideration	
Grades 1 & 2	17	
Grades 3	15	

Source: ISO 2254-1980: International Organization for Standardization

Table 5 Physico-chemical quality requirements/specifications for ground (powdered) clove

Physico-chemical characteristics	Requirements of grades			Test method
Moisture% (m/m) (maxm)	10	10	10	ISO939
Total ash%(m/m) on dry basis (maxm)	7	7	7	ISO928
Acid - insoluble ash %(m/m) on dry basis (maxm)	0.5	0.5	0.5	
Volatile oil ml/100 g on dry basis (min)	16	16	14	
Crude fiber% (m/m) maxm	13	13	13	

Source: 1SO2254–1960 International Organization for Standardization

Storage

Following criteria must be strictly adhered to:

1. Containers with cloves should be stored in covered premises, well protected from sun, rain and excessive heat
2. The store room should be dry, free from objectionable odors and proofed against entry of insects and vermin. Ventilation should be good to ensure enough coolness during summer and sufficiently free of dampness in rainy conditions. It would be advisable to have fumigation facilities in the warehouse
3. Optimum storage conditions are the following:
    1. Moisture 6.5%
    2. Optimum storage temperature 25–28 degree centigrade
    3. Relative humidity 30–40%

Transport

During transport it must be ensured that the product is well protected from rain, excessive heat and intrusion of undesirable odors or vermins.

Defective Cloves

1. Khoker cloves – which have undergone fermentation due to defective drying
2. Headless cloves – those without the ball-shaped unopened flower bud at the top
3. Mother cloves – clove fruits produced after fertilization
4. Extraneous matter: Dirt, stones, stems, pieces of wood and clay particles

    All the above must be discarded.

Composition

The following table (Table 6) contains data of clove composition.

Adulteration

Whole cloves are sometimes adulterated with headless cloves, stems, and other ones described as defective ones earlier. Adulteration is most difficult to identify in clove bud oil. Another adulterant is clove terpenes obtained as by products in the extraction of eugenol from clove oil. Adulteration with synthetic terpinol, dibenzyl ether is best detected by the odor on a blotting paper, after standing for a few days.

Gross adulteration with constituents foreign to clove oil is seldom encountered. Detection of water-soluble adulterants can be done by determining any change in properties before and after washing the oil with saturated salt solution. The most important test in the analysis of clove oil is the determination of total phenols (chiefly eugenol).

Quality Standards

Quality Standards have been prescribed for clove under PFA and ISI and that for clove oil under ISI. Agmark (Indian) Standards have not been prescribed for clove and clove oil.

Packing of Cloves

Cloves are normally packed in double jute sacks with 50–60 kg capacity. Clove buds are purchased in lots of 2.5 tonnes, but, 5 tonnes is the usual minimum quantity for both clove buds and clove stems.

Packaging of Clove Oil

The preferred packaging for clove oils is the lacquer-lined 200 kg or 220 kg capacity stainless steel drums. Polythene containers are used for smaller quantities; but, it is recommended that the oil is not stored in them for long periods. Minimum quantities purchased from stem and bud oils may be as little as 400 kg, but, clove leaf oil is normally imported in minimum lots of 1–2 tonnes.

Table 6 Composition of clove

Property	Content
Moisture	5.4%
Protein	6.3%
Volatile oil	13.2%
Non volatile ether extract (fat)	15.5%
Crude fiber	11.1%
Carbohydrates	57.7%
Mineral matter	5.0%
Ash insoluble in HCl	0.24%
Calcium	0.7%
Phosphorus	0.11%
Iron	0.01%
Sodium	0.25%
Potassium	1.2%
Vitamins mg/100 g:	
Vit B1	0.11
Vit B2	0.04%
Niacin	1.5%
Vit C	80.9
Vit A	175 IU
Calorific value (food energy)	430calories/100 g

Packaging of Clove Oleoresin

The oleoresin is normally packed in plastic pails of 25 kg capacity, but, the size may vary. No minimum quantities are specified by the trade (Smith and Anand 1984).

Processed Products

Clove Bud Oil

The volatile oil derived from the dried buds by steam distillation (yield 16%) contains as its main constituents, tree eugenol (70–90%), eugenol acetate and caryophyllene. Although these substances amount to some 99% of the oil, they are not responsible for the characteristic fresh and almost fruity note of the clove bud oil, which is due to traces of other compounds, the most important being methyl-n-amyl ketone. A special distinction of clove bud oil is that it contains a substantial percentage of eugenol acetate whereas clove-stem oil and clove leaf oil contain only traces of it.

Clove Stem Oil

The chemical composition of clove oil derived from clove stem (yield 4–5%) has not been investigated thoroughly. This is because clove stem oil is not as important as clove bud oil, commercially. Percentage of free eugenol is more compared to clove bud oil. Substances which occur only in traces, which impart the characteristic, almost fruity odor to the bud oil, seem to occur in stem oil in far lesser quantities, or lack entirely, which explains the coarser and "flatter" odor of the stem oil. On the other hand, clove stem oil contains a few constituents which have not yet been observed in bud oil.

Clove Leaf Oil

Clove leaf oil (yield 1–2%) usually contains a somewhat lower percentage of total eugenol compared to clove bud oil, and, eugenol acetate occurs in the leaf oil as in the stem oil, but, only in small quantities. As far as the substances which occur in the stem oil (not in the bud oil) are concerned, namely, sesquiterpene alcohol and naphthalene, only the latter (traces) has been observed in leaf oil. Quality/chemical composition of clove bud oil, Clove stem oil and Clove leaf oil with reference to the aroma quality of clove has been investigated (Gopalakrishnan et al. 1988).

Oil of Mother Cloves

The oil distilled from mother cloves (yield 6.5%) has the following characteristics:

1. Specific gravity: 1.0933
2. Refractive index: 1.5433
3. Phenol content: 88% (eugenol 60%, a solid phenol, having paraffin-like color is present)

Clove Root Oil

It is obtained by steam distillation of clove tree root with a yield of about 6% oil. It is bright yellow in color, when freshly distilled, and in composition, odor, and quality, it compares well with oil from clove buds, contains 85–95% eugenol.

Uses

As Food Flavorant: Clove is very aromatic, has a pleasant flavor and imparts warming qualities. It is widely used in Indian kitchen and blends well with sweet and savory dishes. Cloves, both whole and ground, are widely used in baked goods, cakes, confectionery, chocolate, puddings, desserts, sweet syrups, preserves etc. Clove which has a stimulating effect is widely used in Indian *paan* (the betel leaf and arecanut combination for chewing). In Java, clove is used in the preparation of a special brand of cigarette.

In Medicine: The essential oil in clove has digestive properties, and is both anti septic and anti biotic in treating tooth ache. Externally, it has counter-irritant action. Also, it is an ingredient in many tooth pastes and mouth washes. Because of its anti septic and anti bacterial properties, clove oil finds its use in many pharmaceutical preparations. It has also been found beneficial in blood sugar control, and hence, helps diabetics.

In perfumery and cosmetics: The use of clove oil in perfumes, toilet waters, and soaps of oriental and spicy odor is well established.

Other Uses: The lower priced oil from cloves, in particular the leaf oil, offers an excellent starting material for the isolation of eugenol, which can be converted to isoeugenol and a very high grade vanillin. However, vanillin and eugenol are now synthetically prepared which are cheaper than the eugenol from clove.

References

Chezhiyan et al (1996) Standardization of seed propagation technique for clove and nutmeg. Spice India 9(6):21–22

Gopalakrishnan M, Narayanan CS, Mathews AG (1988) Chemical composition of clove bud-stem and leaf oil. Indian Perfumer 32(2):224–235

Kumar N, Pugalendhi L, Thamburaj S (1992) Clove-an observation on germination of clove seen in Shevroys. Spice India 5(11):7–9

Palani M (1988) Cultivation of clove in South India. Spices News Letter, April 1988 22(4):18–21

Smith AE, Anand N (1984) The U.K. market for cloves, nutmeg, mace, cassia and cinnamon. Market Rep. G184. Tropical Prod. Dev. Institute, London

Mace and Nutmeg

Botanical Name: *Myristica fragrans* Hout
Family: Myristicaceae

Description and Distribution

Mace and Nutmeg are two distinctly different spices produced from fruit of an evergreen aromatic nutmeg tree, which is usually 9–12 m tall, but, sometimes reaches a height of 20 m or more. Mace is the dried reticulated "aril" of the fruit and nutmeg is the dried seed-kernel of the fruit. When the peach or apricot-like looking nutmeg fruit bursts open along its suture, or groove at the center, the mace is seen as an attractive, bright, scarlet-cage, closely enveloping or clothing the hard, thin, black, shining shell of the seed called "nutmeg". At this point the fruit is harvested with a special device. The husk (rind) is broken apart carefully. The mace is skilfully removed from the seed shell, gently pressed flat, dried in partial shade, ad is called the typical "blade of mace"; on drying, the original scarlet color of the mace turns rather pale yellowish brown and becomes brittle. The husk or rind is generally pickled or candied. The nutmeg (kernels) are left within the shell and dried until they rattle when shaken. The very thin shells are then carefully broken and kernels (known as the real nutmeg) are removed, dried and packed for sale.

Nutmeg is a densely foliated evergreen tree growing to a height of about 20 m and more. The tree is normally unisexual, bearing either male or female flowers. However, it is not uncommon to come across bisexual trees. The male flowers are borne in clusters in axils of leaves. The female flowers, also borne in leaf axils, are often solitary. The fruit is a fleshy drupe, spherical or slightly ovoid in shape and pale yellow in color with a longitudinal groove in the center.

The flavor of mace is similar to that of nutmeg, but, is more refined. Mace is much more expensive than nutmeg. Poor mace has usually little aroma and is brittle.

K. P. Nair, *Minor Spices and Condiments*,
https://doi.org/10.1007/978-3-030-82246-0_14

Both the spices are used to flavor a number of foods and beverages and are also used in medicines. Their pleasant aroma is due to the essential oils present in them, which again are similar in organoleptic and physio-chemical properties.

East Indian nuts have the finest aroma. They are ovoid, approximately 2.25–2.75 m long, 1.75–2.25 m in diameter, and, longitudinally, wrinkled. The color is greyish-brown, with network of furrows, of dark veins, in which volatile oil is found.

Myristica fragrans is a native of Moluccas, now cultivated in many countries of both hemispheres – Malaysia, Indonesia, West Indies, Grenada, Sri Lanka and India. In India, it is grown in Tamil Nadu, Kerala, Assam. Preliminary field trials have shown that the Araku Valley (in Andhra Pradesh) and Waynad district in Kerala are well suited for its cultivation.

Area, Production and Exports/Imports

Cultivation of nutmeg in India is mainly confined to Kerala and Karnataka States. No reliable data are available on the actual production of nutmeg or mace. Further, whatever is produced in the country is unable to meet the internal demand, and, consequently to meet it, the only route is to import from neighboring countries like Sri Lanka, Indonesia, Singapore etc.

Crop Management

Climate and Soil Nutmeg thrives well in areas up to 600 m above mean sea level with a warm, humid climate. Well distributed annual rainfall up to 250 cm is ideal for the cop. It can be grown in a variety of soils. The crop cannot tolerate excessively dry or waterlogged conditions. Presence of Iron (Fe) in the soil is beneficial for its good growth.

Propagation by Nursery

Nutmeg is usually propagated through seed. Large-sized "tree-burst" fruits from high yielding trees are harvested or collected as and when they fall to the ground. The fleshy rind on the fruit as well as the mace covering the seed are removed before sowing. The seeds should be sown in specially prepared nursery beds immediately after collection, or within 24 hours as the seeds have only very low viability. If there is delay in sowing, seeds may be kept in trays filled with moist sand.

Nursery beds 100–120 cm wide, and 15–20 cm high and of convenient length are prepared in a cool shady place. A mixture of garden soil and sand in the ratio of 3:1

may be used to prepare these beds. Sand may be spread over the beds to a thickness of 2–3 cm. Seeds are sown 2 cm deep in these beds at a spacing of 12 cm apart and covered with soil. The seed beds may be regularly watered with rose can at least once a day. Normally seeds germinate within 60 days. When young seedlings are at least at 2-leaves stage, and 15–20 cm tall, they can be transplanted into pots/polythene bags. Delayed transplanting damages the roots.

Standardization of Seed Propagation Technique for Nutmeg: Advancements

The following table (Table 1) summarizes results of an investigation on seed grading on germination and growth parameters.

Transplanting of Seedlings in the Field

When 12–18 months old, seedlings are transplanted into the main field. Recommended spacing is 8 m x 8 m. Pits of 60 cubic meters are dug and filled with compost or farm yard manure and top soil. Seedlings are planted carefully in the center of the pits during rainy season. Young plants are provided with artificial shade and irrigated during summer months. As nutmeg requires light shade especially at early stage, banana or glyricidia *(Glyricidia maculata)* plants are planted in between the pits. Though not essential, irrigation is preferred for grown up trees to yield better.

Mixed Cropping Nutmeg can be advantageously grown mixed with coconut and arecanut. When coconut is planted at a spacing not less than 8 m x 8 m, one nutmeg tree can be planted at the center of 4 coconut trees. In arecanut plantation raised with a spacing of 2.7 m × 2.7 m, one nutmeg can be planted at every third row of arecanut so that within the square formed by 4 nutmeg trees, there are 9 arecanut seedlings.

Table 1 Effect of seed grading on germination and growth parameters in nutmeg

Seed Grades	Germination (%)	Plant height after 60 days (cm)	Number of leaves after 60 days
4-5 g	32	3.90	2.8
5-7 g	62	14.20	5.0
7-9 g	71	17.00	7.0
9–11	49	4.20	2.8

Source: Chezhiyan et al. (1996)

Manuring

Nutmeg requires heavy manuring. Organic manure (compost or farm yard manure) can be applied at the rate of 10 kg/plant right during the first year after planting. The quantity should be increased each year so that a well grown 15 year old tree receives about 50 kg/tree. Among synthetic fertilizers, 20 g N, 18 g P, and 50 g K are applied per plant. The synthetic fertilizer dose may be increased gradually until a well grown15 year old tree receives 500 g N, 250 g P, and 1000 g K annually.

In May–June organic manure is applied, in one dose, while the synthetic in two, one along with the organic manure and the second in September–October. The fertilizers are applied in shallow trenches dug all around the tree. As large number of roots grow on the soil surface, care should be exercised while digging the trenches so that the roots are not injured. For trees 15 years and older, trenches are dug about 1.5 meters away from the main trunk of the tree. After applying the fertilizers the trenches must be closed with top soil.

Mulching

Nutmeg is a gross feeder. It is advisable to mulch them, supplemented by application of 25 kg compost/tree/per annum.

Interculture/Weeding

As nutmeg is a shallow rooted tree, no interculture is done which will injure the roots. However, weeding is done periodically.

Plant Protection

Pests No serious pests are seen on the nutmeg tree. However, occasionally, mealy bugs and scale insects attack tender portions of the stem and cause drying of parts by sucking sap. Spraying 0.05% Dimethoate (Rogor) is done to control these pests.

Diseases

Fruit Rot Immature fruits, either split or unsplit, are shed in large numbers. The losses could be severe if timely control is not done. Spraying 1% Bordeaux mixture or Dithane M-45 at 0.5% concentration effectively controls the diseases.

Die-back The disease is characterized by the rotting of leaves and young stems. Control measure include cutting and removing affected parts and applying Bordeaux paste on the cut surface and spraying 1%Bordeaux mixture.

Thread Blight Though not very serious, it is a common disease on nutmeg. The leaves dry and fall. The falling leaves are held attached to branches by means of the fine filamentous fungal threads, and, therefore, the name of the disease "thread blight". Spraying 1% Bordeaux mixture is also effective in controlling the disease.

Constraints in Nutmeg Cultivation

The main hurdle in nutmeg farming is because of its dioecious nature, as the male and female plants are detected only after flowering. The nutmeg tree is slow growing and flowers only about 6–7 years after planting. Occasionally, male trees also bear some fruits, but, only female tree is the main crop yielder. Generally, only 50% of the plants raised turn out to be female. Hence, it is a great disadvantage to the farmers, who have to spend a lot of time, labor and money to realize that 50% of the trees raised which turn out to be are unproductive males (Krishnamoorthy and Rema 1989).

Harvesting and Curing

Under favorable conditions, nutmeg tree starts to bear from the 7th–eighth year after planting, and, attains full bearing capacity by 15–20 years, and thereafter goes on bearing well for 25–30 years. However, in Sri Lanka, one can find even 80–90 old nutmeg plantations to be quite productive.

Curing of nutmeg and mace consists of open sun drying, and its effectiveness depends on the availability of bright sunshine. The quality of the product is not uniform. Therefore, it is essential to modify either the sun drying method or use artificial drying method where possible (Kachru et al. 1989).

The seeds are dried separately for 4–8 weeks either in the sun or in artificial heat drier until the kernel rattles inside the shell which is then broken open carefully with a wooden mallet and the kernel is taken out and suitably packed and stored with or without liming.

Gopalakrishnan et al. (1980) evolved a suitable curing, drying and packing method to retain maximum natural appearance, color and flavor. The seeds after separation of aril were dried in a try-type Kilburn made electrical cross flow drier in a single layer. At 65 degree centigrade air temperature, case hardening was observed which further hindered normal drying process. The drying at 50 degree centigrade required 28 hours duration to reduce moisture content of the seeds from 36% to 9% in comparison to 7 days required in open sun drying. However, sun drying, though slow, dries the seeds uniformly. Artificial drying even at 50 degrees centigrade showed "case hardening" effect after 20 hours of drying. It is, therefore, essential to stimulate natural sun-drying conditions during artificial drying so as to get a good quality and uniformly dried product (Kachru et al 1989).

Yield

A fully grown tree, under normal conditions, may produce 2000–3000 fruits per annum. On an assumption that 50% of the trees in a plantation are females and optimum yielders, about 800 kg of mature nutmeg and 100 kg mace can be harvested per hectare. According to another report, mace is very light in weight. For every 100 kg nutmeg a tree produces only 4 kg mace can be obtained.

Economics of Cultivation

The estimated expenditure in bringing up a hectare of plantation to the bearing stage (eighth year) is approximately INR 20,000 (1 US $ = 75INR). The annual maintenance charge thereafter will be approximately INR 2500/ha. As 800 kg nutmeg and 100 kg mace fetches approximately INR 20,000, net return of INR 17,500/ha can be expected from one hectare.

Post Harvest Technology

Packaging of Mace and Nutmeg

Nutmeg from the West Indies, Grenada ad Indonesia is packed in 50–60 kg capacity double jute sacks. In Sri Lanka, double polypropylene bags are used. Minimum quantities purchased are 2–5 tonne lots, and, maximum 100–200 tonnes. Larger lots have lesser freight charges.

Mace is usually packed in wooden cases of 60 kg capacity each, but, smaller sizes are sometimes shipped. Minimum quantities are for nutmegs. Mace pickings may be pressed into 90 kg bales.

East Indian nutmeg oil is packed in 180–200 kg capacity steel drums (suitably enameled/coated).

The nutmeg oleoresin is normally packed in plastic PVC pails of various capacities, such as, 15, 20 and 25 kg PVC, of which 25 kg is the most popular in the industry.

Grading

Mace Commercial mace consists of flattened lobed pieces, 2–5 cm or more in length, somewhat less in breadth and 1 mm thick. When soaked in water, the lobes swell up and regain their original form. It is dull yellowish red, translucent and brittle. In odor and taste it resembles nutmeg, but is softer and more delicate. Four grades or types of mace are recognized:

1. *Banda Mace* is considered to be the finest. It has a bright orange color and fine aroma
2. *Java Estate Mace* is golden yellow, interspersed with brilliant crimson streaks, line Banda Mace
3. *Siauw Mace* is comparatively inferior in quality
4. *West Indian Mace* is comparatively inferior in quality

Papua Mace, often regarded as the fourth grade of East Indian mace, is derived from *N.gentia*. It contains comparatively little volatile oil and that too of an undesirable turpentine-like aroma. It is unsuitable for distillation.

Banda and "*Penang Mace*" are considered to be of the best in quality. This "true" mace must be differentiated from mace consisting of aril of *Myristica argentica* Warsh known as "Macassar" or "*Papua Mace*" or of aril of *Myristica malabarica* Linn known as *"Bombay Mace*" or "*Wild Mace*". The first tree is native to New Guinea and the second to India.

Being a thin lacy material, the mace is very light in weight and consequently, for every 100 kg of nutmeg, a tree produces only 3–3.5 kg mace. It's quality depends upon its essential oil content.

Mace is available in the market as "whole", "broken" or "ground". It is in great demand in India. During 1971–72 about 78,400 kg of mace valued at more than INR 100, 000 (a huge amount at the time) was imported into India, mostly from East Asian countries. Area under nutmeg/mace is gradually being extended in southern India, especially Kerala and Tamil Nadu.

Nutmeg

The following classification holds for Nutmeg:

1. Large (80 to a lb) which are in great demand
2. Medium (110 to a lb)
3. Small (130 to a lb)
4. There is also a grade "ABCD" un-mixed unassorted.

Out of the above, the first category is in great demand.

Sound Shrivels: These are used for grinding but too expensive for distillation.

Rejections: Considerably lower priced, can be used for distillation.

Broken and Wormy: The cheapest grade, large quantities of which are exported to Europe, especially Hamburg, in Germany, for distillation.

Grinds or Particle Size: Fine grinds are not recommended for both nutmeg and mace as they promote caking. A practical grind for both is the one which passes 95% of the contents in the US Standard sieve no. 25 or Tyler sieve no.24.

International (ISO) Commercial Quality Grades and their Standards Specification. The Commercial International (ISO) categories/grades and their specifications for nutmeg are in the following tables (Tables 2 and 3).

The commercial categories of nutmeg are given in the following table (Table 4).

Note

Whole nutmegs are sold either as a mixture (ABCD) without sorting by size, or by the number of nutmegs to the (British) pound after sizing. For example, "Whole sound nutmeg 80″ (80 superscript please) would indicate nutmegs of category 1, of a size corresponding to 80 to the British pound. The most commonly used size ranges in commerce are: 60–65, 80–85, 100–110, 111–120, 121–130 and 131 to 150 to a pound.

The physico-chemical characteristics of some of the important nutmeg and mace types are given in the following tables (Tables 5 and 6).

The comparative quality/composition of Mace and Nutmeg are given in the following table (Table 7).

Table 2 Commercial categories of Siauw Type ad Banda Type Mace

Category	Commercial designation	Description
1.	Whole Mace (entire)	Whole mace containing more than 5% of about ¼ size broken pieces
2A.	Broken Mace I	Mace containing at least 75% of pieces larger than about ¼ size
B	Broken mace II	Pieces smaller than ¼ size
3.A	Sifting I	Small broken pieces
B	Sifting II	Very small broken pieces

Source: ISO:DIS6577

Table 3 Commercial categories of Grenada Type Mace

Category	Commercial designation	Description
1.	Whole no. 1	Whole mace, pale yellow in color not more than 5% of broken pieces
2.	Broken no 2	Pieces ranging from yellow to dark red in color, but overall yellow of at least ¼ size, with a small proportion of smaller pieces and dust (less than 5%)
3.	Mace pickings	Dust and discolored pieces of mace ranging from yellow to black and containing damaged country mace

Source: ISO:DIS6577-1988

Table 4 Commercial categories of Nutmeg

Category	Commercial designation	Description
1.	"Whole sound nutmegs"	Whole, sound well- formed nutmegs classified as function of their size (see note below
2.	"Sound shrivels"	Shrivelled, but sound nutmegs which are not insect-infested
3.	"BWP" (South East Asia) or defectives (West Indies)	Broken nutmegs or defectives (West Indies) whole nutmegs that have been insect- infested (but without any living insects and practically free from dead insects) they may have been limed

Source: ISO;DIS6577–1988

Table 5 International (ISO) Physico-chemical specifications for Whole Nutmeg (Siauw Type, Banda Type and Grenada Type)

S. No.	Physico-chemical characteristics	Requirements
1.	Moisture content(%) (m/m)(maxm)	10.0
2.	Total ash, %(m/m) on dry basis (maxm)	1.5
3.	Acid-insoluble ash, % (m/m) on dry basis (maxm)	0.5
4.	Volatile oil content, ml/100 g (min)	5.0
5.	Calcium content, expressed as calcium oxide, % (m/m) on dry basis (maxm)	0.35

Source: ISO:DIS6577-1988

Processed Products

Mace Products

Oil of Mace: It resembles nutmeg in odor, flavor, and composition and no distinction is made between them in trade. Like nutmeg oil, mace oil also becomes viscous on storage due to absorption of oxygen. Old mace yields more viscous oil than the fresh one, hence, the importance in judging the freshness of mace oil.

Table 6 Physico-chemical Requirements/Specifications for hole Mac (Siauw Type, Banda Type and Grenada Type)

S.No.	Physico-chemical characteristics	Requirements
1.	Moisture content % (m/m), maxm	10
2.	Total ash% (m/m) on dry basis, maxm	4
3.	Acid-insoluble ash, % (m,m) on dry basis, maxm	0.5
4.	Volatile oil content, m/100 g, min	5.0

Source: Draft ISO: 6577-1988

Table 7 Quality/composition of mace and nutmeg

	Property	Content
Mace	Moisture %	15.9
	Protein %	6.5
	Ether extract %	24.4
	Carbohydrates %	47.8
	Fiber %	3.8
	Mineral matter %	1.6
	Calcium %	0.18
	Phosphorus %	0.10
	Iron mg/100 g	12.6
	Volatile oil %	4–15
	Amylodextrin %	25
	Vit B1	0.37
	Vit B2	0.56
	Niacin	1.2
	Vit C mg/100 g	12
	Vit A IU/100 g	175
Nutmeg		
	Moisture %	14.3
	Protein%	7.5
	Ether extract %	36.4
	Carbohydrates %	28.5
	Fiber%	11.6
	Mineral matter%	1.7
	Calcium%	0.12
	Phosphorus %	0.24
	Iron mg/100 g	4.6
	Volatile oil %	6–16
	Starch%	14.6–24.2
	Pentosans%	2.25
	Furfural%	1.5
	Pectin%	0.5–0.6

Fixed Mace Oil: Mace yields a fat similar to that from nutmeg, but, in a much smaller quantity. A sample of Indian mace gave 26% of red-colored fat (18–20%) after removal of volatile oil on extraction with carbon tetrachloride. It has the following properties:

Specific gravity: 0.984
Refractive index: 1.4850
Acid value: 3.4
Saponific value: 108
Iodine value: 153–157
R.M. value: 7.2
Polenske value: 0.72
Acetal.value: 65–67
Unsapon. Matter: 35%

The characteristics of refined oil, after removal of volatile and resinous matter were as follows:

Specific gravity: 0.9769
Refractive index: 1.4835
Saponic and Iodine value: 118–119

The fixed oil is a semi-drying oil. However, it may not be economical on commercial scale.

The amylodextrin is present in mace in the form of granules, visible under the microscope (5 Mu – please use symbol for Mu). They are compound and irregular in shape with a distinct hilum.

Adulteration of Mace Pure mace, whole or broken, is sometimes adulterated with "Wild Mace" (*M.malabaricum*) of inferior quality. In the case of ground mace, the common adulterants are fernaceous products and ground wild or cheaper mace.

Nutmeg Butter

Nutmeg contains 38–43% of either extractable material which, in addition to glycerides, contains a volatile oil (6–13%), a small quantity of resin, and a substantial proportion of unsaponifiable material. Commercial nutmeg butter, a highly aromatic fat, is obtained from undersized, damaged or worm-eaten kernels which are unfit for sale as spice. The material is ground and cooked or steamed before pressing. A yield of 24–30% is reported. The fat may be obtained by solvent extraction also, but, this process is not usually employed.

Nutmeg butter is soft, solid (melting point 38–51 degree centigrade), yellow or yellowish-red, with the odor and taste of nutmeg. The analytical characteristics of the commercial product vary greatly according to the method of preparation and the proportion of the volatile oil and resin present. The recorded ranges of the values are:

Specific gravity at 20 degree celsius: 0.945–0.960
Refractive index: 1.4662–1.4604
Iodine value:33–65
Acid value: 10–25
Sap. Value:154–190
Unsapn.matter (containing myristin): 8–18%
Volatile oil of bark: The bark of the tree yields volatile oil with the following characteristic:
Specific gravity: 871
Optical rotation: 12.2
Sap.value:14
Ester value after acetylation: 37.5
A variety of" Kino" is obtained from the injuries made on the bark
Volatile Oil from flowers: A volatile oil is obtained from flowers
Stem utilization: The stem contains a tannin-mucilage complex
Edible Food Products from Fruit Rind: Fresh pericarp from ripe fruit (about 80%) contains an acidic (2.3%) astringent juice with an aromatic flavor. Analysis of the fruit rind showed:

1. Moisture 86.8%
2. Protein: 1.0%
3. Ether extract: 0.4%
4. Carbohydrates:11.2%
5. Mineral matter:0.6%
6. Calcium; 0.04%
7. Phosphorus:0.01%
8. Iron: 2 mg/100 g
9. Carotene (as Vitamin A):8 IU/100 g

Fresh husk and rind are a rich source of pectin. A number of delicious edible food products like nutmeg jelly, chutney, preserve, candy, pickles, beverages, jam, mixed fruit jam and canned fruit cocktail (in combination with pine apple, papaya orange and banana) have been developed (Pruthi et al. 1980).

Adulteration of Nutmeg: Pure nutmeg is sometimes adulterated with "false" nutmeg (*M.malabaricum*) or "Bombay nutmegs" which are practically colorless and tasteless. The oil is sometimes adulterated with turpentine. Only an expert analysis can detect these adulterations.

Uses

Nutmeg

Both nutmeg ad mace are used as condiment, particularly in sweet foods, baking etc. In eastern countries they are used more as a drug than as a condiment. Oil of nutmeg or mace is employed as a flavorant. Mace is also similarly used; it is chewed to mask foul breath. Nutmeg butter is used as a mild external stimulant in ointments, hair lotions and plasters and is useful in controlling rheumatism. It has anti bacterial property.

Alcoholic extracts of nutmeg show antibacterial activity. The volatile oil has weedicide properties. It may also be used for imparting scent to soaps, chewing gums etc. It is not produced on a commercial scale.

The pericarp or rind of the ripe fruit is locally used in pickles, also in the preparation of jellies, jams, chutneys etc. Half ripe fruits are candied in Malaysia.

Mace

Both nutmeg and mace are classified as baking spices. They are frequently used in Frankfurter formulae and in recipes for curried meat and other products. Also extensively used in soaps and preserves. For light colored foods, such as pound cakes, cream pies and cream soups, mace is often chosen because of its light orange color.

In India, both are extensively used as drugs and as a condiment due to their valuable medicinal properties. Both are stimulants, carminatives, astringent and aphrodisiacs and can be used in pharmaceutical preparations to control dysentery, stomach ache, flatulence, nausea, vomiting, malaria, rheumatism, sciatica and leprosy. However, nutmeg is used more in the kitchen and in medicine than mace. Their volatile oil is used as a flavorant in liquor, tobacco and dental creams.

References

Chezhiyan N, Ananthan M, Vedamuthu GBP (1996) Standardization of seed preparation techniques in nutmeg and clove. Spice India 9(16):21–22

Gopalakrishnan M, Thomas PP, Bhat AV, Varkey AG, Menon N, Mathew AG (1980) Post-harvest technology of nutmeg. Placrosym 111:189–200

Kachru RP, Srivastava PK, Patel RT (1989) Post-harvest operations for tree spices. Spice India 2(12):13–57

Krishnamoorthy B, Rema J (1989) Grafting – A solution to the sex problem in nutmeg. Spice India 2(6):13–15

Pruthi JS, Bhat AV, Satyavati K, Kutty, Gopalakrishnan M (1980) Chemical composition and development of edible by-products from nutmeg waste (rind). Proc. Sym. on By-products of Food Industries AFST, pp 19–20

Allspice or Pimenta or Pimento

Botanical name: *Pimenta officinalis* Lindl
Syn (1): Pimenta dioica (Linn) Merr (2) *Eugenia pimento* DC
Family: Myrtaceae

Description and Distribution

"Allspice" or "Pimenta" or "Pimento", comprises dried unripe berries from a bushy evergreen tree. The berries are nearly globular, 4–7 mm in diameter, and, when dried, look like oversized pepper berries, with a somewhat rough surface and reddish-brown color. Pimenta is quite different from "Pimento" (*Capsicum annum)* which is a variety of red pepper. The term "Pimento" had its origin in the initial incorrect belief of the early Spanish explorers who thought these allspice berries to be similar to pepper. Then came the English expression "*Jamaica Pepper*" and the German term "*Nelkenpfeffer*". It is now officially recognized as "Pimenta". The name "*Allspice*" is derived from the fact that the spice is said to possess the characteristic blend or composite flavor and aroma of clove, nutmeg, cinnamon and black pepper, all rolled into one. Allspice is available both in whole or ground form. It belongs to the same family as of clove.

Allspice trees are medium-sized growing up to a height of 8–10 m, evergreen, with a slender upright trunk and smooth greyish bark. Two kinds of fruiting trees are noticed, generally designated as male and female trees. Both male and female trees look alike and cannot be identified until flowering, just like in nutmeg.

K. P. Nair, *Minor Spices and Condiments*,
https://doi.org/10.1007/978-3-030-82246-0_15

Distribution

The allspice tree discovered in the Western hemisphere by the early explorers is indigenous to West Indies (possibly, this is the reason why the British called it "Jamaica Pepper"), Central America, Grenada, Guatemala, Mexico, Honduras, which grows semi-wild in Jamaica. In India, it is cultivated in West Bengal, Bihar and Odisha states. It also grows well and fruits well in Karnataka and Kerala States (Waynad district).

Area, Production and Exports/Imports

Although allspice has an international commercial importance, there are no reliable data in India about its production and acreage or export/import statistics. A transaction around US $ 1.5–3.2 million annually, involving around 3000–4000 tonnes would give a rough idea of allspice production and trade; however, this data is not up-to-date.

Crop Management

Soil and Climate

Pimenta can be grown in a wide range of well-drained soils from plains to an altitude of 1000 m above mean sea level with annual rainfall ranging from 15 cm–100 cm. However, at higher and cooler altitudes the rust disease is prevalent.

Propagation

For raising seedlings, only fresh ripe fruits are selected from well developed clusters of regularly fruiting trees. The ripe fruits are soaked in water overnight and seeds are extracted next morning by squeezing the fruit. Sowing should be thin and seeds covered with fine soil. Germination commences in about a fortnight, and continues for about 45 days. The review by Thankamani et al. (1994) gives many details.

Propagation by Seed/Nursery Seedlings

On germination the seeds should be transferred to polythene bags or mud pots filled with potting mixture (farmyard manure and soil) and grown in humid shady conditions. An year after germination, when the seedlings are about 40 cm tall they are transplanted in the main field at a spacing of 6 m.

Viability of Allspice Seedlings

The following table (Table 1) gives details on germination, viability etc.

Propagation by Budding

An auxiliary bud together with its leaf and a small sheath of bark is taken from the selected regular bearing tree, and any wood attached to the insides is removed. This shield is inserted into accurately cut space in the bark of a seedling and secured with a plastic grafting tape, and the whole is covered with a plastic sheet. If successful, the bud will grow in about 6 weeks.

Propagation by "Approach Grafting"

In this method the healthy seedlings are taken to female tree, which has been established in a nursery. The best age of seedings is between 18 and 24 months. The female parent tree is selected based on (1) its form or shape, (2) yield of berries, (3) oil percentage, (4) composition of its berries. Opposing tongues, about 6 cm log, are cut in the seedling stem and a comparable branch of the selected scion tree. The tongues are pressed together to ensure the maximum contact between the two

Table 1 Mean parameters of viability of Allspice seeds

Treatments	Germination (%)	Days to first germination	Days to complete germination
Sowing weeks after collection			
On the day collected	93.33	27	69
1 week thereafter	70.00	34	75
2 weeks thereafter	93.33	23	47
3 weeks thereafter	83.33	20	54
5 weeks thereafter	66.67	30	79
7 weeks thereafter	43.33	40	57
9 weeks thereafter	53.33	28	58
11 weeks thereafter	Nil	Nil	Nil
13 weeks thereafter	Nil	Nil	Nil
15 weeks thereafter	Nil	Nil	Nil
F test	(P = .05)	(P = 0.05)	N.S.
C.D. (P = 0.05)	23.51	9.91	

Source: Devadas and Manmohandas (1988)
Note: C.D. = Critical Difference, P = Probability test, N.S. = Not Significant

cambia and are bound together by a plastic grafting tape. It takes about 3 months for seedlings to be successfully approach-grafted, during which period, they are watered regularly. The stock is then severed and removed from the parent plant, hardened off under bamboo covering, repotted into larger containers and kept in storage for about 3–4 months until the unions are properly healed and are ready for transplanting into the main field.

Since 1968, the approach grafting method has superseded all the other methods of grafting in Jamaica. Growers ae now buying approach grafted pimenta which comes to bear quicker than conventionally raised seedlings.

Top Working

Attempts have also been made to convert young trees to known bearing strains by "top working". The most successful method has been by cutting back all limbs of the tree to 1.5 m height, but, it is now concluded that top working is unlikely to be a commercially successful venture.

Of late, Pimenta has been successfully budded onto bay (*Pimento racemosa*) and *vice-versa*.

Rooting in Allspice by Air Layering

The Indian Institute of Spices Research, (IISR) Calicut, Kerala State, India, has achieved rooting by air layering using suitable hormones. Over 35 clonal selections have been made by this method in Jamaica, while India is still to make a big leap forward in this technique.

Transplanting of Seedings

Seedlings are transplanted at a spacing of 6 m either way in the main field and in poorer (fertility) soils a closer spacing is followed.

Weeding/After Care and Fertilizer Schedule

Young tress or groves are usually fenced off against livestock until well established. Use of chemical weedicides must be done with care to prevent air drift.

After planting, temporary shade, irrigation and weeding are essential to reduce plant causalities (Krishnamoorthy and Rema 1991). The following fertilizer schedule has been followed in Jamaica (Purseglove et al. 1981, Thankamani et al. 1994, Table 2).

Successful Fruiting in Allspice

Allspice (*Pimenta dioica*) generally does not set seed under all agroclimatic regions in India. Recent investigations at the IISR has demonstrated that combined application of IAA (Indole Acetic Acid) and Benzyl adenine (BAP) in the ratio of 50:5 to allspice trees resulted not only in fruiting, but, also in producing big-sized fruits. The fruits contained as many as 16 seeds, as against the normal of 1–2 seeds.

Harvesting and Yield

Male trees flower earlier than female trees and the usual flowering time in India is during March to June (Krishnamoorthy and Rema 1991). From flowering, fruits mature in 3–4 months. For bearing, 5–6 years are taken under good crop management (Purseglove et al. 1981). A well grown tree yields 20–25 kg berries per annum under good crop management (Krishnamoorthy and Rema 1991).

Major Types of Allspice (Pimento)

Jamaican Allspice: Jamaica is the world leader in allspice production and its product commands premium price. While the average oil content in all others vary between 2–4.5%, the Jamaican variety averages 4%. Some authorities feel the Jamaican trees, which are the most cultivated ones today, are also of a strain different from those found on other Caribbean islands and Central America. The uniqueness of Jamaican allspice is that it is hand picked and carefully sun dried.

Table 2 Fertilizer Schedule for Allspice in Jamaica

Years after planting	Fertilizer mixture(g)		Period of fertilizer application
	10:10:10 NPK 15:15:15 or 12:10:18 or 10:10:20 NPK		
1	113	900	February & September
2	170	1130	February & September
3	227	1360	February & September
10	2220[a]	2300	February & September[a]

Source: Purseglove et al. (1981), Thankamani et al. (1994)
[a]Fertlizer dose gradually increased to 2270–2300

Mexican Allspice: They have larger berries, darker in color. The Mexican strain is different from the Jamaican. Flavor is less sweet ad mellow and average oil content is 2.35%.

Honduran Allspice: The berries are similar to those from Guatemala and Jamaica, although they are larger than the latter. Essential oil content is roughly the same as of Mexican type (2.35%).

Plant Protection

Diseases

Leaf Rust: This is caused by *Puccinia psidii* Went. Infection is by urediospores on young growth and also on expanding leaves, inflorescence etc. Spore cycle takes 2 weeks. Severe infection leads to defoliation, and can ultimately lead to death of the tree.

For good control young leaves and inflorescence are sprayed with dithane M-45 or menzate D.

Die back or Canker: Caused by Caratocytis fimbriata Ell & Halst. Symptoms are similar to that caused on other tree crops, the disease starting on one limb as a result of primary infection below the crotch of the tree, following death of the tree in a few months. The disease can be controlled by careful pruning and removal of infected limbs together with the application of white lead paint.

Insect Pests

Termite or White Ants: Most serious pest attacking weak and damaged trees which gain entry through wounds caused to the tree by reckless branch breaking during harvest of berries.

Black Ants: *Crematogaster brevispinosa* Mayr. causes this, which can make harvesting a very painful operation. Ants also spread scale insects between trees. Chlordane spray controls the attack.

Scale Insects: The two principal scale insects of Pimento are (1) *Odonaspis pimentae,* the pimento "armoured scale" and (2) *Laveia primitive* var. *pimentae*- the soft scale insect. Thy attack tree trunks which cause discoloration of bark and bark may shed in irregular-sized pieces instead of the more usual long strips. They can be controlled by white oil spray to which malathion has been added.

White Fly: Unthrifty exposed trees are often attacked by white fly, *Aleurodicus pimentae* Laing. This pest is attacked by the fungus *Aschersonia aleurodis*, particularly in shady and damp places. To control, artificial inoculation is done.

Red - banded Thrips: *Selenothrips nibrocinctus,* the damage appears as silvery-white scale on leaf undersurface, mostly along mid ribs. Badly affected trees may fall can be controlled by malathion or dimethoate spray.

Bagworm Caterpillars (*Oeceticus abboth*) feed on the leaves, sometimes denuding small plants. They can be controlled by a stomach poison.

Fiddler Beetle: These are adult weevils of *Prepodes* and *Pachanaeus* spp. feeding on leaves, while the grubs girdle the roots. Dieldrin spray will control them.

Post Harvest Technology

Grades/Types

As in other spices, the quality of pimento entering the world market, differs according to their geographical origin as well as their oil content. Consumers express preferences for particular types in certain applications. A clear distinction is made between Jamaican pimento and the others produced in Guatemala, Mexico or Honduras. Jamaican variety sets the standards for world market, as it possesses best flavor, aroma, appearance and volatile oil content (4–4.5%). The quality of pimento is affected by several factors, such as, (1) growing region, (2) type of tree, (3) stage of maturity of berries at harvest, (4) method of distillation, (5) freedom from mold and (6) storage conditions.

The ISO physico chemical requirements are given in the following table (Table 3).

The physico-chemical properties for ground product are given in the following table (Table 4).

International Organization for Standardization

Storage of Pimento

Containers of pimento should be stored in well ventilated, dampness-free, free of objectionable odors, store houses.

Table 3 ISO Physico-chemical specifications/requirements for whole Pimento

S.No.	Physico-chemical characteristics	Requirements	Test method
1.	Moisture content % (m/m) (maximum)	12.0	ISO:939
2.	Total ash, % (m/m), on dry basis (maximum)	4.5	ISO:928
3.	Acid-insoluble ash, % (m/m) on dry basis (maximum)	0.4	ISO:930
4.	Volatile oil ml/100 g on dry basis (minimum)	3.5	

Source: International Organization for Standardization (ISO): 937-1980

Table 4 Physico-chemical specifications/requirements for ground Pimento

Physico-chemical characteristics	Requirements	Test method
Non-volatile ether extract,% (m/m)on dry basis, (maximum)	8.5	ISO:1108
Crude fiber, % (m/m), on dry basis (maximum)	27.5	ISO:5498
Volatile oil ml/100 g. on dry basis (minimum)	2.8	

Transport of Pimento

Transport should be in such a condition that the consignment is free from rain, excessive heat, entry of vermins and other objectionable odors, especially if the shipment is through sea.

Packaging and Storage

Packing is similar to Black pepper in gunny bags, internally lined with polythene.

A typical analysis of allspice berries (ground) show the following characteristics (Table 5).

Processed Products

The following products are manufactured from Pimenta:

Pimenta Berry Oil

This is obtained by steam distillation of the crushed berries for about 10 hours, it's yield is 3.3–4.5%. The oil has a yellow to yellowish-red color, darkening with age and possesses the characteristic color and flavor of allspice. It's chemical characteristics are as follows:

Specific gravity at 15 degree Celsius: 1.024–1.055
Refractive Index: 1.525–1.536
Optical rotation: 0.50 (0 small zero superscript please)-5.00 (0 small zero superscript please)
Alcohol: 70%
Eugenol: 65–89%, principal constituent like the clove oil

Table 5 Composition of ground pimento

Property	Content
Moisture	8.8%
Protein	6.0%
Fat (ether extract)	6.6%
Fiber	21.6%
Carbohydrates	52.8%
Total ash	4.2%
Calcium	0.8%
Phosphorus	0.1%
Sodium	0.08%
Potassium	1.1%
Iron	7.5 mg/100 g
Vit C (ascorbic acid)	39.2 mg/100 g
Vit B1 (thiamine)	0.1 mg/100 g
Vit B2 (riboflavin)	0.06 mg/100 g
Niacin	2.9 mg
Vit A	1, 445 IU
Food energy or calorific value	380 calories/100 g

Pimenta Leaf Oil

Dried pimeta leaves on steam distillation yields 0.7–2.9% oil which contains eugenol (65–96%) as its main constituent, has an inferior flavor and odor compared to pimeta berry oil. Used as a chief substitute for expensive berry oil.

Pimenta Oleroresin

Berries are processed, on a small scale by some of the importing countries, to obtain oleoresin, but only little information is available on its manufacture.

As Food Flavorant

Berries are used as condiment, as a flavoring agent in ketchups, soups, sauces, pickles, canned meats, sausages, gravies, fish dishes, pies, puddings, preserves etc.

Use in Medicine

Allspice is used as aromatic, stimulant, digestive and carminative. Powdered fruit is used to control flatulence, dyspepsia, and, diarrhoea, also used as an adjuvant in tonics and purgatives. It's an anodyne against rheumatism and neuralgia.

Berry Oil

There is a good market for berry oil. Oil finds its use in soap manufacture, men's spice – based cosmetics like after shave lotion and cologne. Widely used in food industry. It has also pharmaceutical properties and is anti microbial.

Leaves

Leaves contain tannin and may be used in tanning industry. Oil extracted from leaf (0.7–2.9%) can be used as a cheap substitute for expensive berry oil.

Pimenta Bark and Wood

Pimenta tree wood is used to make walking sticks, umbrella handles etc. Wood from tree trunk is of dark to light salmon color with very firm, hard, close texture and smooth surface.

The above review clearly shows that all parts of the Pimenta tree are commercially utilizable. Hence, it would be a great economic proposition for a hard working farmer.

International organizations conducting research and development in Pimenta/Allspice:

1. Agricultural Research Station, Kade, Ghana
2. Commonwealth Mycological Institute, Kew, Surrey, UK
3. Ministry of Agriculture and Lands, Jamaica
4. Indian Institute of Spices Research, Calicut, Kerala State, India
5. Tropical Products Development Institute, London, UK
6. University of California, Davis, USA

References

Devadas VS, Manmohandas TP (1988) Studies on the viability of Allspice seeds. Indian Cocoa, Arecanut and Spices J 11:99
Krishnamoorthy B, Rema J (1991) Allspice. Spice India 4(10):9–10
Purseglove et al (1981) Spices Vol 1 & 2, 1st edn. Trop. Prod. Inst, London
Thankamani CK, Sivaraman K, Kandiannan K, ad Peter KV (1994) Agronomy of tree spices (clove, nutmeg, cinnamon and allspice). A review. J Spices and Aromatic Crops 3(2):105–123

Curry Leaf

Botanical name: *Murrayya koenigii* (Linn) Sprengel
Family: Rutacae

Description and Distribution

Curry leaves or *curry patta* or *kary patta* (patta in the Indian Hindi language meaning leaf) are extensively used in Indian cookery for centuries. They are used to flavor curries, *sambhar, rasam, chuteys* (south Indian delicacies), hence the plant constitutes one of the major leaf spices of India – an adjunct to Indian curried delicacies.

Curry leaves are derived from a handsome, aromatic, more or less deciduous small shrub, about 0.9 m in height in a Deccan plateau or a small tree up to 5 m in height and 15–40 cm in diameter, growing in the Himalayan region. It is found throughout India and the Andaman islands up to an altitude of 1500 m. It is commonly found in forests, often as gregarious undergrowth, along the foot hills of the Himalayas from the river Ravi to Sikkim and Assam. It is also found growing in Bengal, Madhya Pradesh, Maharashtra, Tamil Nadu, Kerala, Karnataka and Andhra Pradesh. It is extensively cultivated for its aromatic leaves and for its ornamental value throughout India.

Area, Production and Exports

It is cultivated on a plantation scale in Dharwad, Belgaum and Uttar Kannada districts of Karnataka State. Madalageri and Bhagavantragouda (1989) have described how it can be cultivated on a plantation scale and if the right cultivar is planted, 4–5

K. P. Nair, *Minor Spices and Condiments*,
https://doi.org/10.1007/978-3-030-82246-0_16

Table 1 Export of curry leaves from India during 1991

Countries exported to	Quantity Exported (kg)	Value(INR)
Canada	222	8153
UK	4400	73,350
USA	1344	48, 200
Total	5966	129,503

Source: APEDA. 1992. Government of India

harvests can be taken if the cultivar in question is insensitive to the cold winter. A high density bush culture plantation of curry leaf cultivars has been established at the Horticulture Garden of the University of Agricultural Sciences at Dharwad in Karnataka State. Abraham (1991) has reported that in Andhra Pradesh there are about 800 acres in Guntur and Krishna District. No reliable statistics is available in India on its acreage and production.

The following table (Table 1) gives an idea of export of curry leaves from India.

Crop Management

Propagation is by seed which germinates freely under partial shade. A spacing of 4.5–6.09 m has been suggested to plant seedlings. Curry leaf plant does not require much of after care except the routine cultivation practices. It is a hardy perennial tree spice which can be raised in marginal lands easily by small or marginal farmers and can fetch the farmer of INR 7000–10,000 per annum/ha from 4 cuttings/annum.

The plant requires minimal labor and it is practically a risk-free crop. There are no distinct varieties. However, it is claimed by some curry leaf farmers that the red-petioled (leaf stalk) variety has a stronger aroma and this gets preference in the market. For commercial cultivation, it is retained as a short bush of 4.5 m height. The crop can be maintained for at least 25 years in a field (Anon 1989).

Curry Leaf Cultivation in Homelands of Southern Kerala, India

A survey conducted in 60 homelands in Kollam District of southern Kerala State, India, showed that curry leaf occupied a small but definite and significant position among the cultivated crops. It was the "housewife's property" grown purely for domestic needs and as such no scientific cultivation practices were recommended. The following table (Table 2) brings out some interesting observations.

Table 2 Comparison of curry leaf cultivation as a homeland crop and a commercial crop

Particulars	Curry leaf Home gardens	Commercial curry crop
Type of plant	Both narrow and broad leaved	Broad leaved not petioled type
Planting material used	Root suckers	Seedlings
Manuring	Only at initial establishment	Applied after each pruning/harvest
Application of chemical fertilizers	Not applied	Rarely applied
Pests and disease control	No chemical control	Regular prophylactic sprays of low residual effect
Harvesting	Daily clipping of 5–10 compound leaves. No harvest at the first and second year	First pruning when plant is 8–10 months old, 30 cm high. Thereafter 3–4 harvests/year
Economic benefit	Only for domestic use	Income of INR 16–40/plant/ annum

Source: Abraham (1991)

Raising Nursery

Good and healthy seeds selected from vigorously growing plants should be used for raising nursery. Young seedlings 3 months old, about 20 cm high, are planted in small pits 15 cm x 15 cm x 15 cm at a spacing of 1 m x 1 m.

Manuring

Application of organic manure once in 2 years is recommended. No chemical fertilizers are applied. Curry leaf plant requires 150 g N/annum/plant, and 25 g each of P and K (Goudra and Madalageri 1989).

Harvesting

In about an year after planting the plant grows to a height of 1–1.5 m and harvesting should commence. In one year about 4 cuttings can be had. The plant is pruned to a height of 30 cm after each harvest.

Weeding

After each harvest the field should be thoroughly weeded and copiously irrigated. Weeding is the most important after care cultural practice for curry leaf, annually 30 weedings are recommended.

Prophylactic Measures

Nuvacron is sprayed (100 ml/60 l water /0.2 ha) at fortnightly intervals. Harvesting should follow spraying after a gap of 15 days to prevent any residual effect on leaves.

Cost of Cultivation

About INR 600 is required for 0.2 ha of curry leaf plantation. Income would be about INR 7000.

Plant Protection

A number of insect pests severely damage the crop (Gowda et al. 1990, Jalaluddin et al. 1990). The various pests which attack the crop and their control measures are discussed below:

Citrus psylla (*Diaphorina citri* Kuw. Family Psyllidae or Hemiptera). This insect is widely distributed in India and is a major pest. Damage is caused by both nymphs and adults, clustering around terminal shoots and suck sap leading to their drying up.

It can be biologically controlled by predators like *Coccinella septempunctata, C.repanda* Th. It can be chemically controlled by insecticides like malathion.

Black fly: *Aleurocanthus woglumi* Ashley. Family Aleyrodidae or Hemiptera. A serious pest damage, same as in the case of *D.citri*. Can be biologically controlled by *Prospaltella divergens.*

Scale insects: *Anoidiella orientalis* Newst. Family:Diaspididae.

Hemptera. It is circular, yellowish-brown in color, and about 2 mm in diameter. The scale is polyphagous and prefers dry weather and infests entire tree, attacking leaves, twigs and branches; heavily infested tree sheds leaves and dries up. There are several generations within an year. There are several exo and endo parasites of this insect.

Biological control: Nymphs of the insect are parasitized by *Anagyrus pseudococci* Gir.

Leaf Roller: *Tonica zizyphi* Staint; Family: Oecophoridae or Lepidoptera. This is a common pest attacking the plant in wet months of May–June. The pest is contained by spraying contact insecticide. Biologial control: *Pristomerus.* sp. and *Brachymeria euploeae* parasitize the larvae of the insect.

Tortoise beetle: *Silana farinosa*. Family: Cassididae or Coleoptera. The damage is caused by grubs which feed on epidermis of the plant at night or tissues of green leaves and sekletonize them. It can be controlled by dusting the crop with 5% malathion or 4% endosulfan.

Citrus butterfly: *Papilio demoleus* Linn. Family: Papilionidae or Lepidoptera. The larvae feed on leaves and defoliate the trees. Biological control: The larvae are attacked by a yellow wasp *Polistes herebrens* and preying mantis *Crebrator gemmatus*. Chemical control by a contact herbicide with a short-lived residual effect.

Bark borer: *Indarbela tetraonis* M. Family: Arbelidae or Lepidptera. It has an all-India presence and often a serious pest in southern India. Spraying the bark with a residual insecticide during lean season or at the time of pruning the shrub.

Citrus leaf miner: *Phyllocnistis citrella* Stainton. Family: Phyllocnistidae or Lepidoptera. Leaf is heavily damaged due to the mining of the insect. At times even the young shoots are attacked and the whole tree suffers. *Scotolinx quadristriata* Rao & Rema parasitize the insect. It is, rather, difficult to control the pest. Prompt pruning and destroying infested leaves must be done to control effectively the pest.

Leaf weevil: *Myllocerus discolor, M.viridanus*. Family: Curculioide or Coleoptera. Adult beetle feeds on leaf edges, causing serious damage. Dusting with less persistent insecticide controls the pest.

Hairy Caterpillar: Yellow in color, belong to the family Lymantridae, found in clusters at early instars, skeletonize leaves.

Coreid Bugs: Two species of the bugs are found to infest tender shoots. Details are given by Jalaluddin et al. 1990.

Termites: Severe attacks are seen in red soils (Gowda et al. 1990) and a spray of 17 ml dimethoate with 25 ml of dicofol in 10 l of water once in 3 months controls the pest (Gowda et al. 1990).

Extreme caution must be exercised while using chemical insecticides because the leaves are directly consumed.

Diseases

A disease caused by *Cychinorosporum* sp. reported from Dharwad of Karnataka State is found to be far more serious than any other pest attack.

Diseases

White Sap Rot: Caused by *Fomes pectinatus* Klotszch.

Collar Rot of Seedlings: Caused by *Rhizoctonia (Corticium) sobani* Kuhn, reported from Nagpur district of Maharashtra State.

Another precaution must be to avoid spraying sulfur compounds since leaf shedding has been observed consequent to that (Krishnarajan 1991).

Post Harvest Technology

Packing ad Storage.

The leaves are dried in shade or dehydrated in a suitable dehydrator and packed in containers, kept closed. The following table (Table 3) gives details of yield.

The following table (Table 4) gives the composition of curry leaf.

The leaves are a fair source of Vit A, also a rich source of calcium; but, due to the presence of oxalic acid in high concentration (total oxalates 1.35%, soluble oxalates 1.15%) its nutritional availability is adversely affected.

The following table (Table 5) gives data collated by CFTRI, Mysore.

Leaves contain free amino acids, such as, asparagine, glycine, serine, aspartic acid, glutamic acid, threonine, alanine, proline, tyrosine, tryptophan, aminobutyric acid, phenylalanine, leucine, isoleucine and traces of ornithine, lysine, arginine and histidine.

Processed Products

Volatile Oil

Fresh leaves on steam distillation under pressure (6.3272 kg/cm^2) yield 2.6% of a volatile oil which may find use as a fixative for heavy type soap perfume; distillation at ordinary pressure gives only a poor oil yield, while distilling at superheated steam (temperature 220 degree Celsius) yields a dark-brown colored foul smelling oil.

Essential oil obtained by simple water distillation has the following characteristics:

1. Specific gravity (at 25degree Celsius): 0.8589
2. Optical rotation: −36.14
3. Acid value: 1.17
4. Saponification value: 6.18
5. Saponic value after acetylation: 52.81

The total oil yield is 1%, light yellow colored, spicy, aromatic note.

Table 3 Curry leaf yield details

Years	Yield (kg)/ha)
First year	100–150
Second year	750
Third year	750
Fourth year	1000
Fifth year onwards	1500–2000

Source: Krishnarajan (1991) Note: Leaves can be harvested once in two-and-a half months to three months interval

Table 4 Composition of curry leaf

Property	Content
Moisture	66.3%
Protein	6.1%
Fat (ether extract)	1.0%
Carbohydrates	16.0%
Fiber	6.4%
Mineral matter	4.2%
Calcium	810 mg
Phosphorus	600 mg
Iron	3.1 mg
Carotene (as Vit A)	12,600 IU
Nicotinic aid	2.3 mg
Vit C	4.Mg/100 g
Thiamine and riboflavin	Absent

Table 5 The chemical composition of fresh green leaves at maturity, namely, tender, medium mature and mature on moisture-free basis

Property	Tender leaf	Medium mature	Mature
Protein	5.44%	6.44%	7.19%
Fat	3.3	4.74	6.15
Sugars	14.9%	17.9%	18.9%
Starch	11.4%	14.2%	14.6%
Crude Fiber	6.0%	6.2%	6.2%
Volatile oil	0.82%	0.55%	0.48%
Acetone extractive oleoresin	1.6%	1.4%	1.3%
Total ash (mineral matter)	12.54%	12.7%	13.1%
Acid insoluble ash	1%	1.3%	1.35%

Dehydrated Curry Leaves

Recent research at CFTRI, Mysore has shown that vacuum-shelf dried leaves retained a better green color than did the sun-dried ones; dried leaves yielded more essential oil indicating that excess moisture content plays a role in the release of the oil from the leaf, primarily retentive, in nature.

As Food Flavorant: Curry leaves have been used as a natural flavorant in the preparation of curries, chutneys etc. in South India, for centuries.

In Indigenous Medicine: Leaves, bark and root of the curry leaf plant are used in the preparation of indigenous medicine, as tonic, stomachic, stimulant and carminative. An infusion of the roasted leaves is used to stop vomiting. Externally, they are used to heal skin eruptions and bites of poisonous animals. Green tender leaves are eaten to control dysentery. The leaves, fruits, bark and roots are reported to contain a glucoside called "Koenigin". Aqueous extracts of the leaves when administered parenterally to female guineapigs not only raises their phagocytic index, but, also mobilize a greater number of leucocytes to take part in phagocystosis. However, the effect does not last long. The root juice is taken to relieve kidney pain in humans.

In Perfumes and Cosmetics: The essential oil obtained from leaves by steam distillation under high pressure (as indicated earlier in this chapter) can be used as a fixative in heavy type of soap perfume.

Fruits: Fruit is edible, yield 0.76% of a yellow volatile oil with peppery taste accompanied by an agreeable sensation of coolness on the tongue.

Wood: The wood (43.50 lb./cft) is greyish white, hard, even and close – grained and durable, and, can be used in the manufacture of agricultural implements.

References

Abraham VA (1991) Curry leaf cultivation in Andhra Pradesh. Spice India 4(11):13–14

Anon (1989) Curry leaf a risk-free crop. Spice India 2(7):29–30

Goudra KHB, Madalageri BB (1989) Effect of nitrogen nutrition on yield and yield components of curry leaf. Ind Cocoa, Arecanut & Spices J 12:66

Gowda PHR, Shiva Shankar KT, Gowda JVN (1990) Uses and cultivation of curry leaf. Spice India 3(2):7

Jalaluddin SM, Chinniah C, Thirumurthy S, Chiinaswami KN (1990) Insect pests of curry leaf. Spice India 3(12):7–11

Krishnarajan J (1991) Curry leaf cultivation. Spice India 4(3):5–6

Madalageri BB, Bhagavanantagoudra KH (1989) Karibevu Labhadayaka yale tarakari (Kannanada language). Samyukta Karnataka 57(105):4

Tamarind

Botanical name: *Tamarindus indica* Linn.
Family: Leguminosae

Description and Distribution

Tamarind is one of the most common, yet, most important tree spice of India. Almost every part of the tree has a use, but, the most useful are the fruits or pods. India is one of the most important and largest countries producing tamarind in the world, which is chiefly grown in Madhya Pradesh, Andhra Pradesh, Tamil Nadu ad Karnataka. It is estimated that about 250,000 tonnes of tamarind pulp is marketed within India annually for internal consumption. Its bark is brownish grey, longitudinally and horizontally fissured; leaves are paripinnate, up to 15 cm long and leaflets are generally 10–20 pairs, sub sessile, oblong, 8–30 mm. Flowers are small, yellowish with pink stripes in lax, few flowered racemes at the ends of the branchlets. Pods are 7.5–20 cm long, about 2.5 cm broad and 1 cm thick, more or less constricted between seeds, slightly curved, brownish-colored and scurfy. Seeds are 3–12, obovate-oblong, compressed, with a shallow, oblong pit on each side of the flat face, and are 1.5 cm long, smooth, dark brown shining. The seeds are contained in loculi, enveloped by a tough, leathery membrane, the so-called endocarp. Outside the endocarp is the light-brownish, red, sweetish, acidic, edible pulp traversed by a number of branched, ligneous strands. The outermost covering of the pod is fragile and easily separable.

The tree has usually a short and thick trunk, seldom straight. It has a handsome, dense, spreading crown of feathery foliage and is prized for ornamental purposes. Though an evergreen tree, there is abundant leaf fall and new leaves appear in March–April. In very dry regions it sometimes becomes leafless during hot summer. Flowers appear from April–June and the pods ripen in winter.

K. P. Nair, *Minor Spices and Condiments*,
https://doi.org/10.1007/978-3-030-82246-0_17

Area, Production and Exports

India produces about 250,000 tonnes of pulp annually and most of it is consumed internally. Some quantity is exported both in fresh and dried forms. There is scope to export “instant tamarind concentrate” and” tamarind powder”.

Crop Management

Soil and Climate

The tamarind tree is not exacting, inasmuch as soil fertility is concerned. It has become naturalized even in rocky land. However, it thrives best in deep alluvial soil. Tamarind is suited to semi tropical climate, with low rainfall. It can come up even in saline soil.

Varieties

Tamarind is generally classified on the basis of color and taste of tamarind pulp as red and sweet types. In 1991, the Tamil Nadu Agricultural University’s Horticultural College and Research Institute released a variety PKM1 for commercial scale cultivation. It has a pulp recovery of 39%. The pulp is also very sweet. Pods are less fibrous. It yields 250 kg/tree in the ninth year after planting. It is generally propagated through seeds (Suresh et al. 1992).

Propagation

The tree is easily propagated by seeds, though it can also be propagated through “cuttings” or by “budding”. Propagation by means of a “gootee” has been suggested to ensure replication of desirable parental characters. The seeds are directly sown along the ploughed or hoed lines, and germination generally commences in 5–10 days. For transplanting, the seeds should be sown in April in raised nursery beds of light and porous soil and transferred to the site during the first rainy season, before the tap root becomes too long. Some success has also been obtained by transplanting during the second rains with the stem and taproot pruned down to 5 cm and 22–23 cm respectively. Transplanting potted seedling in pits, 1 m x 1 m x1m at a spacing of 3 m x 3 m has also been successful.

Patch Budding for Quicker Multiplication on Large Scale

Pathak et al. (1992) based on their investigations have recommended "patch budding" for quicker multiplication of tamarind tree on a large scale.

Growth/Development of Seedlings

The tree grows to a long height, attains a large size, but, growth retards after seedling stage. Under favorable conditions, early growth can be quite rapid. With good after care, a height of 60 cm can be attained in the first season itself. Pits may be filled with a mixture of farm yard manure, top soil and 50 g BHC (BHC 10%) dust. Seedlings or grafts should be planted in the center of the pits. While planting grafts, care should be taken to keep graft -union above ground level. Stack young seedlings to protect them from strong winds.

Irrigation

Regular watering of the trees is essential until the seedlings establish, and, thereafter irrigate as and when required.

Manuring

Generally, tamarind is not manured, but, application of farm yard manure (FYM) helps in boosting yield. Seedlings may be applied 25 kg FYM, 0.5 kg urea, 1 kg superphosphate and 0.5 kg muriate of potash in the first year, which helps in good growth and subsequent yield.

After Cultivation

The following steps must be ensured for good tree growth and yield:

1. Remove rootstock sprouts as and when they appear
2. Remove dried and dead plant parts
3. Intercrops can be raised during the first 3 years to compensate the establishment cost, for which vegetables, groundnut, sesame and sorghum can be grown as intercrops depending on the availability of water and soil fertility

Disease Control

The tamarind tree is adversely affected by a number of root diseases, such as "sap rot" caused by *Xylaria euglossa* Fr., "brownish sap rot" caused by *Polyporous calcuttensis* Bose, and "white rot" caused by *Trametes floccosa* Bres. Among the other diseases, *Meliola tamarindi* Syd.causes the "sooty mold". *Pholiota gollani* P.Henn has been recorded on the trunk of the living trees. *Polystictus sarbadhikarii* Bose on the dead trunk, and different species of *Hypoxylon* Bull. Ex Fr. on the decorticated wood, branches and the bark have been recorded. A bacterial "leaf spot" has been recorded on trees near Pune in Maharashtra State, India.

Insect Pests

Several insects attack the trees and when the product is stored. Among them, the beetles of larvae of *Lasioderma serricore* Fabr. *Pachymerus gronagra* Fabr. and *Calandra linearis* Herbst bore the seeds; larve of *Tribolium castaneum* Herbst, *Argyroploce illipedia* Butler and *Aspidiotus* spp. are other pests which destroy the fruits ad seeds.

Nematodes

Xiphinema citri Siddiqi and *Longidorus elongatus* (de Man) Thorne & Swanger are said to be associated with the roots of the tamarind tree.

The Horticultural College and Research Institute of The Tamil Nadu Agricultural University, Coimbatore, India, has recommended the following plant protection measures to control the tamarind diseases and pest attack (Suresh et al. 1992):

1. Spray 1% nuvacron to control mealy bugs and scales at the time of fruiting
2. "Storage Beetle" can be controlled by spraying 1% endosulfan at the time of fruiting, as its infestation starts in the tree itself
3. Spray 0.1% karathane to control powdery mildew
4. There is one important physiological problem "Anikkai"; investgations are being undertaken by the above mentioned institute

Harvesting

When the tree is about 13–14 years of age, it begins to bear fruits and continues to bear for more than 60 years. The pods should be allowed to ripen on the tree itself until the outer shell is dry and can be separated from the pulp without adherence.

Harvesting is done by merely shaking the branches and should not be beaten down with sticks as the process will injure the blossoms and the buds of future leaves. The pods that are left on the tree are gradually blown off later and picked up from the ground.

Yield

The tamarind tree flowers during December–January and the fruits are ready for harvest in April–May. A good fully grown tree yields about 180–225 kg of fruits per season. If maintained well, 100–150 kg of processed (dried) tamarind can be obtained from one tree from the eighth year onwards from planting (Suresh et al. 1992). On an average, the pod is composed of 55% pulp, 34% seed and 11% shell and fiber. India is perhaps the principal producer of tamarind in the world and the estimated production several years ago was 250,000 tonnes per annum.

Post Harvest Technology

Preparation for the Market

The fruit-pulp is the chief acidulant agent for souring curries, sauces, chutneys, and certain beverages throughout the greater part of India. It is a common article of faith in trade and is preserved and stored for marketing in a number of ways. Most commonly, the pulp is freed from fiber and seed is mixed with about 10% salt and beaten down with mallets, so as to exclude air. It is then packed into gunny bags, lined with palm-leaf matting, introducing only small lots at a time. Pulp with fiber and seed is also quite common in the market.

In another process, the salted pulp is trodden into a mass and made into balls, which are exposed to the sun and dew for about a week. For better preservation, the balls may be steamed for a short time before being exposed to the sun and dew. When first prepared, the pulp s light brownish-red, but, it darkens with time during storage.

Composition

Tamarind Pulp

The following table gives details of the composition of tamarind pulp (Table 1). The ripe fruit, on an average, comprises about 55% of tamarind pulp, 33% of seeds and about 12% of fiber.

Table 1 Composition of tamarind pulp

Property	Content
Moisture	18.2%
Free acid (tartaric)	9.8%
Combined acid	6.7%
Total sugars as invert sugars	38.2%
Protein	2.8%
Pectin	2.8%
Cellulosic residue (Fiber)	19.4%
Mineral matter	2.8%
Calcium	0.17%
Phosphorus	0.11%
Iron	0.111%
Calorific value	83 calories/100 g
Vit A	100 IU/100 g
Niacin	0.2 mg/100 g

Of the reducing sugars present, 70% is glucose and 30 is fruit sugar (fructose). Only a trace of sucrose or cane sugar is present. The pectin present in the pulp is of good quality, having a jelly grade of 180–200.

Tamarind Concentrate

1. Total soluble solids: 65–70%
2. Tartaric acid: 12–13%
3. Ash: 2%
4. Ash insoluble in dilute HCl: 32% of total ash

Tamarind Powder

Tamarind powder has also been developed by CFTRI, Mysore, Karnataka State, India.

Grading

The Agmark authorities have laid down separate specifications for (a) seedless tamarind, light-red to tinged – brown pulp, (b) seedless tamarind, pulp black color, and (c) tamarind pulp with seeds.

Table 2 Comparative characteristics of red and common type pulp

Red pulp	Common type pulp
Moisture: 20.1%	18.2%
Free tartaric acid: 6.6%	9.8%
Combined tartaric acid: 11.4%	6.7%
Invert sugars: 36.4%	38.2%
Pectin: 4.4%	2.4%

There are considerable variations in the acid and sugar contents in tamarind pulp obtained from different sources. In some samples the pulp is red colored due to the presence of pigment chrysanthemom. The so-called red colored pulp is sweeter than the common type, evidently because it has lower content of free acids. The variety generally fetches higher price and is preferred for making preserves. An analysis of the cleaned pulp red and common type pulp gives the following values respectively (Table 2).

Processed Products

Several novel products like "Tamarind Concentrate", "Tamarind Powder", "Beverages", integrated process for the manufacture of pectin, tartaric acid and ethanol (ethyl alcohol) from tamarind pulp have been developed by CFTRI, Mysore. Details can be obtained by logging on to www.central food technological research institute, mysore.com

Uses

As acid and food flavorant: Practically all parts of tamarind fruit can be used in one or another manner. Ripe fruit or tamarind pulp has an important role in the preparation of several Indian dishes. Details can be obtained from the website referred to earlier.

Medicinal Properties/Uses: According to the ancient Indian Ayurvedic system of medicine, the ripe fruit is appetizing, laxative, healing tonic for the heart, anthelmintic, cures *vata* and *kapha* (Ayurvedic system of medicine), heals wounds and fractures. Ripe fruit is also a refrigerant and digestive and is useful in diseases caused by deranged bile, such as, burning of the body and intoxication of liquors. A gargle of tamarind water cures sore throat.

Fluorosis caused by drinking excessively fluoride contaminated water (>3 ppm) can be controlled/cured by drinking *rasam* or *sambharam* (both widely popular vegetarian Indian drinks, the former prepared using curry powder as a base and the latter from buttermilk, both spiked with ripe tamarind fruit concentrate).

Tamarind Seed/Kernel: The chief use of the seed lies in the manufacture of textile sizing powder. It is widely used in sizing jute yarn and some cotton yarns. It is only half as costly as regular starch. About 3000 tonnes of tamarind seed powder is annually used in the jute industry.

Among the other uses suggested for the kernel powder are as "creaming agents" for rubber industry (latex manufacture). Also used as a soil stabilizer and as a substitute for pectin because of its jellying property, although, structurally and chemically, it is a "polyose" and is quite different from both pectin and starch.

Leaves: The leaves also find a number of medicinal uses. They are applied to reduce inflammatory swellings and ringworm infection etc.; and a number of such diseases.

Flowers: They are acrid, sweet, sour, tasty, appetizing; cure "*vata*" and "*kapha*", urinary discharges and bad sweat odor.

Unripe Fruit: It has sour taste and is indigestible, astringent to bowels, cures "*vata*", causes biliousness, cough and hematological problems.

Bark: It is used topically for loss of sensation in paralysis. The ash is given for urinary discharges and venereal diseases such gonorrhoea. The bark has astringent and tonic properties. It heals ulcers.

Seeds: The seeds are astringent, aphrodisiac, useful in giddiness and vertigo. They are applied externally in liver complaints and inflammations.

References

Pathak RK, Ohja CM, Dwivedi R (1992) Adopt patch-budding for quicker multiplication in Tamarind. Spice India 5(1):7

Suresh J, Sundaran V, Ponnuswamy V, Irulappan I (1992) Tamarind – a tree for wasteland. Spice India 5(12):2–3

Asafoetida or Asafetida

Botanical name: *Ferula asafoetida* Linn.
Family: Umbeliferae

Description and Distribution

Asafoetid or Asafoetida is the dried latex or oleo-gum or gum oleoresin, exuded as a very thick and sticky and pasty sap or latex from the living underground rhizome or rootstock or tap-root of several species of *Ferula* (3 of which grow in India, mainly *F.narthex* in Kashmir. These species are the perennial herbs (1–1.5m high), belonging to genus *Ferula*, which are the sources of the oleo-gum resins, used as condiment as well in medicines. The species are distributed from the Mediterranean region to Central Asia.

Many people have a wrong notion that asafetida is cultivated in India. Its main growing areas are the eastern parts of Iran and the western parts of Afghanistan. In no other part of the word is it grown commercially. Some unscrupulous traders/ farmers sell seedlings of *Gardenia gummifera*, a handsome shrub, as asafoetida plants, as its exudation has an odor akin to that of asafoetida. But, this plant can never yield asafoetida of commerce (George 1995).

There are about 60 species of the genus of *Ferula* which are widely distributed in three main regions (1) Central Asia (2) Europe and (3) North Africa. *Ferula* grows from perennial rootstocks and attains annually a height of 1–3m. The main regions afford the various forms of asafoetida, galbanum and sumba. Some of the commercially important plants have been reviewed (Raghavan et al. 1974), details of which are summarized in the following table (Table 1).

K. P. Nair, *Minor Spices and Condiments*,
https://doi.org/10.1007/978-3-030-82246-0_18

Table 1 Commercial plants of asafoetida grown in the world

Species	Region where grown	Remarks
F.aliacea Boiss	Iran	Gum is used as an intestinal antiseptic and carminative and also to control hysteria and epilepsy
F.asafoetida Linn.	Kashmir, Punjab, Iran, West Afghanistan, Africa	Resin is amorphous and reddish brown, Gum resin is used to flavor sauces and also as expectorant, laxative, and antispasmodic
F.communis Linn.	North Africa	Source of a gum known as "Ammoniac of Morocco" Used medicinally in Europe
F.foetida (Bunge) Reget	Southern Turkey, Iran, Afghanistan	Same use as that of F.asafoetida
F.ferulago Linn.	South European countries	Used medicinally in Europe
F.galbaniflua Boiss. et Buhse	N.W. India, Iran, Turkey	Source of galbanum, a gum resinous exudation, orange brown to brownish black. Used medicinally in Europe
F.hermonis Boiss	–	Used medicinally in Syria
F.jaeschkeana Vatke	Kashmir, Turkey	Source of a gum resin
F.marmaria Ash.et Taub	N. Africa	Source of a gum known as "Ammoniac Cyrenaica"
F.narthex Bois.	W.Tibet, Kashmir	Used as a spice for its flavor
F.orientalis Linn	–	Used medicinally in middle East
F.persica Wild	Iran	Gum (sagapenum) is sold in cakes
F.rubricaulis Boss	–	Used medicinally in Iran
F.schair Bosz	Turkey	Probably a source of galbanum. Used medicinally in Europe
F.sumbul Hook.f.	Mountains to the south-east of Samarkhand	Root has a musk-like scent, used as stimulant and tonic for nervous disorders
F.szowitziana	Central Asia	Source of sagapenresin, has a scent of galbanum
F.tingitanaLinn. (Syn.F.sancta Boiss)	Syria, N.Africa	Source of North African and Moroccan ammoniac

Source: Raghavan et al. (1974)

Area, Production and Trade

No official data on area and production of asafoetida are, yet, available in India. Even when some production is there, it is inadequate to meet the internal need. It is estimated that as much as 3000 tonnes of compounded asafoetida is manufactured and sold in India. Yet, India is the largest importer.

Method of Gum-Resin Collection

Ferula plants from which asafoetida of commerce is extracted, have massive tap roots and massive carrot-shaped roots, 12–15cm in diameter at the crown when they are 4–5 years old. The rhizomes produce two types of plants, one of which is the male plant, which produces inflorescence and the other, the female plant, which produces the oleo-gum or asafoetida. In March-April just prior to flowering, the upper part of the living rhizome/root (about 5–6 cm) is laid bare and the stem is cut off close to the crown. The exposed surface is covered by a dome-shaped structure, 15–20cm high, called the *khora,* made of twigs and clay and which has an opening to the northern direction. A milky juice exudes from the cut surface. After some days, the exudate is scraped off and a fresh slice of the root is cut, when more latex exudes, sometimes the resin is removed along with the slice. The collection of the resin and the slicing of the root are repeated until exudation ceases (about 3 months after the first cut). The resin is sometimes collected from successive incisions, made of the junction of the stem or rhizome and the tap roots. With 3 incisions, some plants have been reported to yield about 1 kg or more of gum resin.

Crop Management

There are two types of Asafoetida, (1) *Hing Kabuli Safaid* (white asafoetida) and (2) *Hing Lal* (red asafoeida).

Asafoetida is acrid, and bitter in taste, and emits a very strong, pungent, disagreeable alliaceous odor, due to the presence of sulfur compounds. Hence, the common name abroad – "Devil's Dung".

Quality Evaluation

The quality and flavor of asafoetida are affected by a number of factors, such as, (1) part of the plant from which it is prepared, (2) the season of collection (3) the method of preparation, (4) the degree of adulteration ad (5) the nature of adulterants used.

Composition

The following table (Table 2) gives the composition of Asafoetida

The resin consists chiefly of asaresinotennol, free or combined with ferulic acid. Umbelliferone seems to be present in the combined form.

Analysis of the samples obtained from the market of "Hing", "Hingra", and, "Bandhani Hing" or compound asafoetida collected from the Mysore market, in Karnataka State, by CFTRI showed the following analysis:

Table 2 Composition of asafoetida

Property	Content
Resin	40–60%
Gum	2.5%
Volatile oil	10–17%
Ash	10–15%

Table 3 The quality specifications of commercial asafoetida

Quality attribute	Hing	Hingra	Bandhani Hing
Total ash% (max)	15.0	20.0	10.0
Ash insoluble in HCl% (max)	2.5	8.0	1.5
Alcoholic extract% (with 90% Alcohol)	12.0(max)	50.0(max)	5.0(min)
Starch% (max)	1.0	1.0	–

1. Total ash: 4.4–44.3%
2. Alcohol insoluble matter: 20.8–90%
3. Alcohol soluble matter: 28.3–40.9%

The following table (Table 3) gives data on the quality specifications of commercial asafoetida.

Adulteration

Commercial asafoetida is grossly adulterated with gum arabic, other gum-resins, rosin gypsum, red clay, chalk, barley or wheat flour, slices of potatoes etc. Different grades (Hingra, Hudda Hing, Hira Hing) varying in price are sold in the market.

Processed Products

Oil of Asafoeida

The oil of asafoetida is obtained by steam distillation of gum resin. It's yield can vary from 3–20%. It has the following physico-chemical properties:

1. Specific gravity at 20 degree celsius: 0.906–0.973
2. Optical rotation at 20 degree celsius: −90 (0 small zero, superscript please) 18
3. Refractive index at 20 degree celsius: 1.493–1.518
4. Sulfur conten: 15.3–29%

The chief constituent of the oil is secondary butyl propenyl disulfide. The remaining constituents are other disulfides, a trisulfide, pinene, another terpene, and an unidentified compound. The disagreeable odor of the oil is reported to be mainly due to the disulfide (C11H20 S2 – 11, 20,2 subscripts please). The oil has no commercial importance. The flavoring and pharmaceutical industries chiefly employ alcoholic tinctures of the gum-resin.

Compounded Asafoetida Powder and Tablets: A Simple CFTRI Process

Since pure asafoetida is not preferred for use in view of its strong flavor/odor, it is sold in compounded form. The compounded asafoetida types manufactured and marketed in India is in mass/lump form are reported to have fungal growth and are non uniform in quality. The available market samples of compounded asafoetida invariably contain high moisture (up to 25%), thereby being susceptible to mold; they harden on storage, and thus become difficult to handle. CFTRI, Mysore, has developed a process to form compounded asafoetida into attractive table briket forms. The compounded asafoetida powder/tablet is uniform in quality and convenient to use., for both culinary and flavoring purposes. Owing to its optimum moisture content, it is free flowing (powder)/free from lumping or hardening (tablets) and is also free from mold growth on storage. The volatile oil in the product is fixed at 2%, which is the desirable level.

Details of CFTRI Process

Equal quantities of starch and gum are mixed in a blender. The pasty mass of Afghan asafoetida is soaked in required quantity of water and this slurry is added to the above and mixed well. More gum is added and the mixing continued in the blender. The compounded asafoetida is powdered in a mill and packed as such, or after making tablets. Details of daily production are as follows:

Raw Materials Needed for 200 kg Batch

1. Asafoetida: 40 kg
2. Starch: 40 kg
3. Gum arabica (powdered): 120 kg

The main raw materials required are gum arabica, starch and asafoetida. While the former two are available in the country, the last one has to be imported from Afghanistan.

Safety Considerations
Since asafoetida is lachryniatory in nature, care should be taken to protect the eyes.

Uses

As Food Flavorant: Asafoetida is extensively used in India to flavor curries, sauces, and pickles, in conjunction with onion and garlic. In Iran, the natives rub asafoetida on warmed plates prior to placing meat on them. Besides, the large cabbage-like tops of the plants are relished raw by the natives.

In Medicine: Asafoetida is a stimulant, carminative and antispasmodic. It is used as a uterine tonic. Medicinally, it stimulates intestinal and respiratory tracts and the nervous system. It is useful in asthma, whooping cough and chronic bronchitis. Also, it is often used in veterinary medicine.

Recent investigations have shown that asafoetida has the following properties: (1) its oil has antibiotoc properties and inhibits growth of microbes in intestines and (2) as it has sedative properties, it can be recommended in cardiac diseases

In the Indian Ayurvedic system of medicine, it is a constituent of the well-known "*Hingashataka*".

It is important to realize that the precise scientific manner in which asafoetida helps in medicine is still to be thoroughly investigated, though, many empirical observations, as the ones cited above, have been observed.

Some Important Concluding Observations

The following points are of importance:

1. Asafoetida is still a wild crop. The producing countries have not systematically investigated the economics of its cultivation, and, its commercial potential, in particular, in India. This needs to be addressed
2. High yielding varieties offer immense potential for increasing production/yield. This needs to be investigated systematically. The fact that, by and large, asafoetida is not susceptible to many other problems of diseases and pests, specially insect pets, a systematic effort must be made to look at its production potential
3. A thorough field survey in growing regions, especially in India, is called for, prior to R&D efforts
4. A comprehensive study is to be carried out on post-harvest technology of asafoetida, with the primary focus on pre- empting adulteration
5. Attempts should be made to introduce species, such as, *F.foetida, F.rubricaulis.*
6. India being the largest consumer of asafoetida, early efforts to grow the product commercially in the country should be initiated(George 1995)

References

George CK (1995) A glimpse on asafoetida. Spice India 8(8):2–5

Raghavan B et al (1974) Asafoetida. II. Chemical composition and physico-chemical properties. Flavour Ind 5(7):179–181

Vanilla

Botanical name: *Vanilla fragrans* (Salisbury) Ames.
Syn. *Vanilla planifolia* Adrews

Description and Distribution

The vanilla pods or "sticks of commerce" are the cured fruits and beans of the climbing orchid *Vanilla fragrans* and *Vanilla planifolia*. It is a native of the Atlantic Coast from Mexico to Brazil. Vanilla cultivation had spread to other countries after the discovery of America. The important vanilla producing countries are Madagascar, Mexico, Tahiti, Malagasy Republic, Comoro, Reunion, and Indonesia. Vanilla constitutes one of the most important flavors used in food, perfumery and pharmaceutical industries.

Though vanilla was introduced into India in 1835, yet, not much of a headway has been made in its cultivation across the country, despite its universal popularity as a natural flavorant for use in numerous sweet dishes, notably pudding, chocolate, confectionery, and ice cream industries, within the country, apart from its use in perfumey and pharmaceuticals. Ironically, to meet the internal demand, vanilla worth INR 120,000–400,000 is being imported into India, annually, in addition to the huge quantities of synthetic vanillin and vanilla essence, pointing to the crucial fact that it would be but in the larger economic interest of India that vanilla cultivation is taken up on a huge scale in the country. The data in the following table (Table 1) substantiates this argument.

K. P. Nair, *Minor Spices and Condiments*,
https://doi.org/10.1007/978-3-030-82246-0_19

Table 1 Details of vanilla import into India in 1981–1982

Vanilla products imported	Quantity (kg)	Value in INR (million)
Vanilla	1200	0.12
Vanillin	24,616	1.12
Ethyl vanillin	6.904	0.51
Iso-eugenol (used partly in vanillin manufacture)	5,018	0.32
Total	37,738	2.07

Commercial Forms of Vanilla

Vanilla is commercially known as "Burbon vanilla", "Mexican vanilla", "Indonesian vanilla", and, "Sychelles vanilla". The following 4 commercial forms are well established in the international trade:

1. Vanilla pods, consisting of whole pods which may be split
2. Cut vanilla, consisting of parts of pods, split or not, and deliberately cut or broken
3. Vanilla in bulk, consisting of vanilla in pods and cut vanilla
4. Vanilla powder, obtained by grinding vanilla pods, without additives and after drying

Area, Production and Trade

According the International Trade Center (ITC), Geneva report (1982), about 1600–3000 tonnes of vanilla valued at US$ 45–55 million, enter the world trade each year. This must have increased vastly over the years. No figures on world production have been indicated. Madagascar is the principal supplier, meeting about 75% of world demand, followed by Comoros and Indonesia with 10% each, and, the Reunion, French Polynesia and Mexico supplying about 5%. India does not figure anywhere in these statistics. Efforts are being made to increase the acreage under vanilla in Kerala State, (West Coast), Nilgiris (Tamil Nadu) and to some extent in other states. Only very few cultivators have taken up its cultivation on a commercial scale because vanilla requires hand pollination unlike other crops. Research on vanilla is being conducted at the Indian Council of Agricultural Research, New Delhi, India, controlled Research Station at Ambalavayal in Kerala State, since 1960.

Kumar et al. (1992) have indicated, based on the meteorological and agroclimatic conditions, that the entire Eastern Ghat in Tamil Nadu between 800 m and 1500 m above mean sea level is suitable for vanilla cultivation.

The following table (Table 2) gives data of world area and production of vanilla (1982) by FAO.

In 1988 Madagascar supplied about 50% of the total world supply, while Indonesia shared about 25% and the rest by other countries (George 1989a,b).

Table 2 World area and production (1982)

Country	Area (ha)	Production (tonnes)	Average yield (kg/ha)
Madagascar	8500	1200	141.18
Mexico	347	20	57.64
Reunion	800	27	33.75
French Polynesia	100	5	50.00
Comoros	N.A.	170	N.A.
Quadelope	35	10	28.57
World total including other countries	9782	1432	62.23 (Average)

Note: N.A. Not available

Varieties

"Bourbon" variety from Madagascar, the Comoros and Reunion is highly flavored and preferred, and is thus, the most popular variety. It is marked in three qualities/grades: First, Second and Third, which are based on bean length (between 20 and 25 cm), aroma, flavor, color, consistency/texture, and, absence of defects. Mexico also produces vanilla of good quality, but, its exports are limited.

Future Prospects

The high prevailing price of vanilla should provide an incentive to new growers/producers to enter the world market. Indian growers should take up its cultivation and processing (curing etc.) early and seriously. It can be raised as an intercrop in coffee, cardamom, coconut and arecanut plantations, in Kerala State, India.

Major Constraints in Natural Vanilla Production

The continued increase in vanilla price has attracted many countries to embark on natural vanilla production. However, the emergence of synthetic vanillin, which now accounts for 90–95% of vanilla flavors, has emerged as the main constraint. While the consumers still prefer the natural flavor of the vanilla bean, the price advantage of vanillin, which makes the natural vanilla thrice more expensive than the synthetic vanillin, has emerged as the biggest constraint against natural vanilla production. Hence, there is a race between natural vanilla and vanillin. Globally about 12,000 tonnes of vanillin is produced, while only 2000 tonnes of natural vanilla is produced. As vanilla beans give only 2% of natural extract, this works to just about 40 tonnes of natural vanilla flavor (George 1989b).

A definite Future for Natural Vanilla

There is a widespread awakening, the world over, by the consumers to opt for naturally produced products, primarily in the agricultural sector. Either organic farming, or chemically free end products, the message is clear: people prefer the organically produced products away from the chemically centered ones, and under this ambit comes the synthetic vanillin. It can then be concluded that consumers still prefer the natural vanilla because of its mellowed and well-rounded flavor compared to the rough and incomplete aroma and taste of synthetic vanillin, hence, to offset the price disadvantage of the former, more efforts have to be focused on evolving better varieties and also improve agronomic practices, both of which will immensely contribute to enhance the productivity of natural vanilla.

Crop Management

Vanilla thrives well up to an elevation of about 1066.8 m above mean sea level, under hot, moist, tropical climate with adequate and well distributed rainfall. Arid conditions and violent winds are detrimental to good growth of the vanilla plant.

Soil
Well drained and rich (high humus content) soil, loamy in texture is most suited for vanilla. Depth of soil is not important as the vanilla plant is a shallow-rooted one. However, shade is very critical. In full sunlight, leaves turn yellowish and burn. But, under dense shade, the leaves turn etiolated and spindly. Shady situations, which permit checkered sunlight to filter through, are most suitable for vanilla plant.

Land preparation
Land selected to establish a vanillery (vanilla plantation) should be cleared of undergrowth, retaining only the top shade. If the shade is very dense, the trees should be pruned to allow sunlight to filter through. If sufficient shade is not available, the shade trees should be planted first, and, only when they have established should the vanilla planting follow. Vanilla can be successfully planted in rubber, cardamom, plantations and orchards as an intercrop, and since it requires only minimal amount of cultivation, it will not, in anyway, affect the productivity of the main crop.

Providing Support
The vanilla vines require a support for climbing. In their natural habitat, the vines twine upwards round the trunks of the trees. As the vines grow very high, picking of beans is a hazardous job; hence, their upward growth is restricted by training them on small shrubs, dead-wood posts, or concrete posts. The most commonly used support tree used in the West Coast of Kerala State is *Plumeria alba*, but, *Erythrina lithosperma, Glyricidia maculata, and Indigofera teysmanii* can also be used. In all

these cases, the trees should be pruned and kept to a height of 1.219–1.514 m. Deadwood-posts require frequent replacement, hence, concrete ones could be preferred, though the cost factor must be taken into consideration.

Spacing
Plants are spaced 9 feet within the rows and the rows are 6 feet apart. Support should be provided at these spacings before planting. In the case of live supports, they should be planted well in advance so that they may get well established before planting time.

Planting
Both rooted and unrooted cuttings can be used for planting directly in the field, but, the unrooted cuttings take longer time to bear. Cuttings from any part of the vine, except the tender portion just below the growing point, may be used as planting material. The growth of vines, especially during the pre-bearing period, is considerably influenced by the length of the cuttings, greater the vigor and earlier is the flowering and bean set. The cuttings are first planted in a well-prepared nursery bed with 1or 2 nodes below the soil. Sprouting starts 3 months later and then these can be planted in the main field, about 6 months after they are planted in the nursery bed. Pits, 0.3 m x 0.3 m x 0.3 m should be dug near the supports. The rooted cuttings should be placed in these pits and filled with top soil and well pressed. The cuttings should be gently tied with the supports. Small basins should be made around each plant which should be well mulched with green leaves or leaf mold. If sufficient rainfall is not received, the vines should be watered regularly. The vines grow and cling to the support with aerial roots produced at the nodes.

Two or three supports in a row are connected with a pole put across them horizontally, and when vines are 4–5 feet high, they are coiled round these poles. As the vines grow, more coils are made so that the vines are not allowed to train away from the supports. Poles may be of bamboo or of any strong wood so that they may not break with the weight of the vines. Concrete posts should be provided with 2 or 3 stands of wires.

Manuring

Cattle or any other animal manure is not applied to vanilla as these will have a very deleterious effect on the vines. However, plenty of organic leaf manure can be applied. These should be applied to the base of the vines. Different parts of the vanilla plant have been found to contain large amounts of K and Ca. Hence, the fertilizers carrying these elements should only be applied to those soils which lack them. The best course would be to go by the methods described in the book "Intelligent Soil Management for Sustainable Agriculture – The Nutrient Buffer Power Concept"(Springer, Nair 2019).

After-Care

Vanilla needs only very little after care. Weeding should be limited and minimal, as the plant is shallow rooted and weeding can inadvertently injure the roots. Mulching with fresh leaves, like that of *Glyricidia maculata,* 2–3 times in 1 year is necessary. Irrigation is essential during summer months, especially during the first 2 years after planting.

Flowering and Natural Pollination

The vines flower in the fourth year after planting. The inflorescences are produced on the leaf axils and each inflorescence consists of as many as 15–20 flowers. The tip of the main inflorescence axis continues to grow forming new flowers in succession. The older flowers are at the base and the youngest at the growing point.

Because of the structural peculiarity of the flower, that is, covering the anther by a sort of hood or anther cap, and the stigma by rostellum, natural pollination is difficult. In the countries of vanilla origin, Mexico, Madagascar etc., natural pollination is carried out with the bees of the genus *Melipona* and humming birds. But, in the countries of its introduction, such as India, pollination is carried out by human hand, since the natural pollinators are not present.

Hand Pollination

The following three methods are currently in vogue:

1. Using a small pointed stick of tooth pick size, sharpened at both ends, to transfer the pollen from pollen sac to the stigma
2. Using a thin fat stick to press the rostellum upward and pressing the anther cap with the thumb to dislodge pollen mass on the sticky stigma
3. Using cotton buds to transfer pollen mass from anther cap to stigma.

The second method is commonly employed and the bean set is much higher with this method than with the other two.

A new method of hand pollination has been perfected at the Regional Research Station, Mudigere, Karnataka State. Details are given by Shadakshri et al (1996). One hundred per cent pod set was obtained by this method.

Fertilization occurs in about 24 hours, after pollination. The success or failure of the hand pollination can be judged from the falling away of floral parts. The floral parts remain attached to the ovary after fertilization has been effected. If not, these fall by the third day.

Pod Development

The beans grow quickly in the initial stages and later it slows down. They mature in about 8–9 months. The growth will be much reduced if the floral parts at the tip of the beans fall down during the course of their growth. The bean is ready for harvest when the tip turns yellow.

Tissue Culture for Vanilla Production

The Department of Biotechnology of the Government of India has a very ambitious project on this aspect to cover 150 hectares in India, in three States (Kerala 60 hectares, Tamil Nadu 30 hectares and Karnataka 60 hectares). The Spices Board in Kochi, Kerala State, will monitor the project. Under the scheme, the Spices Board will give INR 4000 per hectare in four yearly instalments. An input incentive of INR 1500 will also be given. Besides the above, planting materials of vanilla (both tissue cultured and rooted cuttings) will be provided to the famers at subsidized costs, as an incentive to propagate extensive vanilla cultivation in India (Anon 1995).

Yield

Though vanilla begins to yield from the fourth year onwards, the maximum yield is obtained at the seventh year. On aveage, 75 beans can be harvested per vine and the total yield from one acre (800 vines) will be 60,000 beans. A number of factors, agronomic, agroclimatic and varietal, decide the ultimate yield. A good vanillery can yield 500–800 kg of cured beans per annum, though much lower yields have been reported from India and Seychelles.

Plant Protection

Diseases

Root Rot: Caused by *Fusarium batalatis* var. *vanillae* and reported from Madagascar and Puerto Rico.

Wilt Disease: It is reported from India and caused by *Fusarium oxysporum* var.*vanillae.*

Anthracnose reported from Madagascar, Comoros, West Indies, Seychelles and Tahiti caused by *Calospora vanillae*.

Apart from the above, there are a few more fungal diseases, but, they are only of minor significance. Vanilla is free from diseases of viral origin. Among the non-pathogenic diseases, or physiological disorders, drying of shoot tip, due to drought, and drying and cracking of stem and leaf tissues on account of continuous exposure to sunlight, are important.

Pests

There are a number of minor pests which inflict minor damages to the different parts of the plant. Among them the following ae to be noted:

1. Trioza bug (*Trioza litsea*) producing holes on buds and flowers
2. Emerald bug (*Nezaria smaragula*) which suck sap of flower buds and stalks, reported from Reunion
3. Á black weevil (*Diorymerelles* sp.) which attack the tip of vanilla shoots
4. Another weevil (*Perissodores nuficollis*) which cause destruction of the vines noticed in Puerto Rico and Madagascar.

Post-harvest Technology

Curing

Free vanillin is not present in the green beans when they are harvested and the characteristic vanilla aroma due to vanillin will also be absent. This is developed as a result of an enzyme action at the time of curing the beans. Curing is a process of alternately sweating and drying the beans until they lose most of the moisture. The following curing process gives a satisfactory product.

The beans after harvest are allowed to shrivel for 3–4 days. They are then immersed in hot water (60 degree Celsius) for about a minute and spread out on a woolen blanket for sun drying. When the beans become too hot to hold in the hand the blanket is folded over them and kept for the rest of the day. At night, the beans are kept in sweating boxes lined with a blanket. The next day, the beans are again sundried as earlier. This process is repeated for about 8–10 days depending on the weather. By this time, the beans lose a major portion of moisture in them and attain a dark chocolate-brown color. The vanilla aroma may also have developed. When properly cured, the beans will be supple and mellow enough to be twisted round the fingers without rupturing. The beans are then sorted out, according to their length, tied into bundles and stored in air-tight containers.

Natural Mechanism of Flavor (Vanillin) Generation During Curing

Vanllin is the most important constituent of vanilla flavor, whose content varies according to the place where the beans are grown, from 1.5% in Mexico to 2.7% in Java. Beans produced in India have been found to contain 2.2%.

Different methods of curing vanilla after harvest are being followed in different countries, as described by Theodose (1973). Of the various processes tried in Indian research stations, researching vanilla, curing vanilla by the Mexican process has been found to be the best under Waynad conditions in the State of Kerala, both for their attractive appearance and high aroma or vanillin content (2.3–2.9%).

Enzymic Generation of Vanillin

During the curing process, vanilla pods get the flavor as the result of the naturally induced enzymatic action of glucosidase on the precursor glucovanillin factors, affecting the formation of vanillin and sugar. Vanillin aroma is the dominant flavor characteristic of vanilla. Climatic conditions, timing of the harvest, and the content of sweating of the pods during curing, are some of the important factors that determine vanillin content and quality of pods.

Mexican Vanilla

The original Mexican process for curing beans involves repeated sweating and drying in the sun. The sweating is accomplished by packing the warm beans into boxes and covering them with mats. The temperature thus maintained encourages enzymatic reactions accompanied by slight fermentation, which develops the flavor of the beans. The process takes, in all, from harvesting to packing, approximately 8 months and is generally believed to produce finest quality of vanilla flavor.

Mexican beans are marketed in 7 grades varying from beans of top quality to the most inferior and "Cuts". Prime beans are the best. They are without blemishes and chocolate brown to black in color. Their aroma is clean and delicate. The beans vary in length between 6.5 and 9.5 inches. Moisture content varies from 20% to 40%.

"Cuts" consist mainly of beans which have been damaged by mildew during the curing process and damaged parts, cut away. "Prime cuts" are usually considered as good as whole beans and are classified as "Shorts" or "Longs". Ordinary cuts are dry broken bans or poorer or underdeveloped flavor from the early gatherings. Whole beans are sold either loose in bulk or made into bundles of 50–90 kg each, tied at both ends and in the center. These are then packed into tin boxes, lined with wax paper to contain up to 39–40 kg each. The tins are soldered and shipped in wooden boxes, 3 tins/box.

The Mexican vanilla, which was once he most sought after, being the original and of the finest quality, has almost disappeared from international trade currently due to huge production decline (Theodose 1973).

Bourbon Vanilla: Vanilla from Madagascar, Reunion and Comoros is known as Bourbon vanilla. In Madagascar, the major vanilla producer, a slightly different and shorter process taking 6–7 months is adopted. In this method, the harvested beans are first placed in hot water (approximately 70 degrees Celsius) for 15–20 seconds and then subjected to a sweating and sun drying process similar to the Mexican one. The method used to cure Bourbon vanilla is said to yield a somewhat stronger and harsher flavor than that obtained in the case of Mexican process.

Bourbon vanilla are classified into 6 grades as follows:

1. Prime
2. Firsts
3. Seconds
4. Thirds
5. Fourths
6. Splits

Prime grades are the finest/perfect, fully mature, smooth, broad, and very oily. They are rich in chocolate flavor with a moisture content of about 37%. Firsts are a degree lower in appearance. Second cuts are sweet flavored and rather inclined towards hardness. They are close in quality to firsts and have a moisture content of about 25.5%. Thirds have good flavor, but surface is blemished. Their moisture content is about 13%.

Fourths are smaller than thirds and are of lighter color and rich in resin content but, low in vanillin. Moisture content is about 16%. Splits are reddish-brown, woody, shriveled and dry. They are high in vanillin and low in resin.

To make a bundle, the beans are tied in the middle, the bundle being packed in tins lined with grease-proof paper, contain 8–10 kg each. The tins are packed at 6 per case.

Indonesian vanilla: Vanilla produced in Indonesia is cured by exporters. No standard method is adopted by them for curing the beans. According to one method beans are killed in hot water (80 degree Celsius) by dipping them for 20 seconds. In a second method hot water (63 degree Celsius) is used and beans are dipped for 2 minutes. Sun drying, sweating, shading, drying and conditioning also vary from one processor to another in sequence, duration and number of times repeated.

To process, beans of different maturity are used. Maturity of harvested beans varies from 5–6 months to over-ripe stage. Immature as well as over ripe beans will have only 15% moisture as against the permissible 20–25% level for extraction purposes and 30–35% for household purposes after curing.

Vanillin is prepared from either whole or cut beans and classified as follows:

Grade 1: Whole beans will have a minimum of 11 cm length.
Grades 2 and 3: Wil have a minimum of 8 cm length.

Minimum vanillin content will vary for Grade 1 at 2.25%.
For Grade 2 will vary at 1.5.
For Grade 3 will vary at 1%.

They are packed in clean tins/cartons/plastic packets with net weight of 15 kg and every two or more packs put in wooden crates/carton boxes (Theodose 1973, George 1989a).

Yield of Cured Beans

About 0.5 kg of cured beans can be obtained from 3.5 kg to 4 kg of fresh green vanilla beans.

Mechanized Method of Curing and Modification of the Vanilla Preparation Method:

In Madagascar, which supplies 75–80% of world vanilla requirements, Theodose (1973) and his associates were busy evolving modified method of preparation and mechanized curing method and mechanical hot-air drying of cut beans (2–3 m long) after sorting, scalding, autoclaving on a commercial scale. An American Company specializing in the production and marketing of vanilla extracts has approved this modified (unconventional) and mechanized process. The economic viability of this new technique on a commercial scale may revolutionize the vanilla processing industry.

Vanilla Bean Packing and Storage

Even after completion of curing, vanilla beans often continue to develop a desirable aroma and flavor, if stored in suitable conditions, and thus, it is quite usual for the vanilla extract manufacturers to store beans for a period after import. Beans are stored in open containers at a temperature of 10 degree Celsius at low humidity (Memory 1956). Moisture content may be between 25% and 30%, preferred for extraction, and storage temperature may be raised to 15–21 degree Celsius, without detrimental effects on flavor quality, if some moisture loss is desired in the beans. Beans cured to proper moisture level as required under the national/international standards are packed in friction-top tins.

Vanilla Pods

Vanilla pods of the same length should be put in packets and then should be put in clean, water tight containers of the material that will have no adverse effect on the product, for instance, tin boxes.

Each of these elementary containers, the contents of which are homogenous, constitute a lot; consignment is constituted of either a homogenous lot or by several lots belonging to different categories.

Cut Vanilla

Cut vanilla should be put in packets of pods of the same length, when they are sufficiently long, and in bulk, when they cannot be put in bundles. They should then be placed in clean, sound and water tight containers of the material which will have no adverse effect on the product.

Cut vanilla should be uniform, from the botanical point of view.

Vanilla in Bulk

Vanila in bulk should be put in clean, sound and water tight containers of the material which will have no adverse effect on the product. If glass containers are used, the words "Glass Fragile" should be indicated on each container to be dispatched. If possible, the year of harvest should also be indicated which will show the freshness of the product.

Vanilla for Retail Trade

Marking should be in conformity with the rules and regulations of the country in which the vanilla is to be sold.

Quality Attributes of Vanilla

The most important quality attributes of cured vanilla beans for grading are:

1. Length of the beans
2. Aroma/Flavor
3. Color

4. Flexibility
5. Luster
6. Free from blemishes
7. Free from mildew disease
8. Free from insect infestation

The best quality pods are 17–25 cm long, highly aromatic, free from mildew, blemishes and insect infestation, dark brown in color, fleshy, supple, and somewhat oily in appearance. It may be cautioned here that the quality of vanilla beans should not be judged only by their vanillin content, since frequently the very finest quality beans show much less vanillin content. The well-cured beans thus graded are packed in air-tight tin containers which can be kept for a long time.

Some Precautions

Some precautions for producing better quality vanilla beans are:

1. Stop harvesting immature/vanilla beans since they yield poor quality vanillin. Harvest when beans at optimum quality, that is, only when blossom-end is yellow
2. Grade harvested beans according to maturity, such as, blossom-end not yellow, just yellow, blossom-end brown, dark brown, and split and process grade separately
3. Process beans by adopting method followed for producing Bourbon vanilla
4. In all stages of processing good beans, split and broken ones should be discarded
5. Prior to filing beans in the boxes, they should be cleaned thoroughly. Check properly the inside of the box
6. Carefully monitor stock against insect infestation, fungal infection and other microbial contamination periodically
7. Control effectively rat menace, if any.
8. Wax paper should be properly spread and kept in the container so that the beans do not come in contact with tin, so that aroma is better preserved
9. The metallic sheet used to make tin boxes and the cardboard to make cartons should be new ad conforming to the quality standards
10. While loading and unloading, care must be taken to prevent damage to tin or carton
11. Spray chemicals inside the processing unit and its surroundings periodically to control insect infestation
12. While spraying chemicals, adequate precaution should be in place not to let the beans come in contact with the sprays, which can be ensured by covering beans and utensils properly covered with plastic sheets (George 1989a)

The following table (Table 3) gives the composition of whole vanilla.

Table 3 Composition of whole vanilla

Property	Content
Moisture	25.85–30.93%
Protein	2.56–4.87%
Fatty oil	4.68–6.74%
Volatile oil	0–0.64%
Nitrogen-free extract	30.35–32.90%
Carbohydrates	7.1–9.1%
Fiber	15.27–19.6%
Ash	4.5–4.7%
Vanillin	1.48–2.90%
Resins	1.5–2.6%
Calcium	19.7%
Potassium	16.2%
Sodium	6.7%
Phosphorus	9.5%
Iron	0.3%

Adulteration

Vanilla beans contain methyl vanillin but ethyl vanillin is synthetic and more aromatic than the former. The choice between the two depends on the kind of preparation for which the product is used and the end use. Ethavan and Vanaldol are the trade names of ethyl vanillin.

Synthetic vanillin can be used as a substitute for the real natural one, for cheap flavoring, such as, vanillin solution and vanillin sugar. Coumarin, a compound with a vanilla-like flavor and taste can used as a cheap substitute for true vanilla. This product has recently been proved toxic and so should be totally avoided/banned against use.

The appearance and odor of the spent (solvent treated) vanilla beans is reported to be improved by treating them with benzoic acid. In the event of suspicion, a simple test can be conducted by heating crystals over a flame which will emit irritating odor due to the presence of benzoic acid, and, this is the signal for its avoidance totally.

Processed Products

Vanilla extracts/Essences

Vanilla extract is in great demand in the US, while cured beans in Europe. The vanilla flavor can be extracted with alcohol. Color of the extract depends on the strength of the alcohol used, duration of extraction, and presence of glycerin. Dark colored extract is obtained from dry beans and the presence of glycerin depends on the color of the extract. Extract can be stored in stainless steel, aluminum or glass containers, and ageing for about 30 days improves its aroma. Wooden containers should be complexly avoided. Natural vanilla extract is about twenty times more expensive than the synthetic one.

Vanilla sugar
Vanilla extract is mixed with sugar and the powder is called vanilla sugar or powdered vanilla.

Vanilla Oleoresin
Prepared by solvent extraction of cut vanilla beans.

Vanilla Powder, "Vanilla Absolute" and Vanilla Tincture

These are used in perfumery, and have been produced and marketed in US and some other countries.

Uses

As Food Flavorant and in Perfumery

Today, vanilla is the most favored flavorant for sweetened foods, the world over. The list is endless, where vanilla sugar is used, in soft drinks, cakes, chocolate, ice cream, liquor, to name just a few. About 50% of vanilla is used in the ice cream industry, alone.

Deodorants, perfumes, odor fixatives, and, as a masking agent in pharmaceutical and vitamin preparations are others where vanilla is used. It is a fragrantly tenacious ingredient in perfumery.

After 1970, the greatest use for technical grade vanillin (98% pure, less desirable for food flavoring) is a chemical intermediate in the production of synthetics of a number of pharmaceutical products. Foaming in lubricating oils can be prevented by vanillin. There are many other such examples.

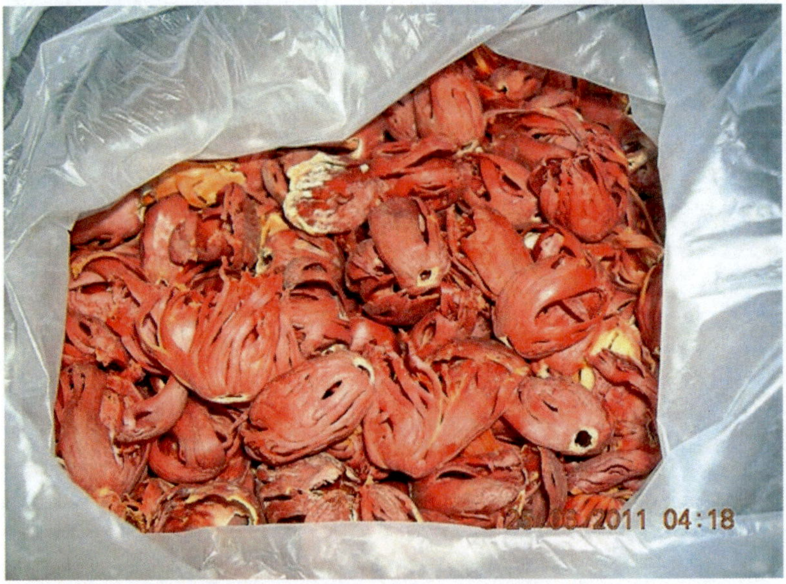

RRLT - NC DRIER (Model 101)

References

Anon (1995) Spice India 8(6)

George CK (1989a) Quality of vanilla beans. Spice India 2(10):3–4

George CK (1989b) International trade in vanilla. Spice India 2(12):3–4

Kumar N, Thamburaj S, Prakasan V (1992) Vanilla-possibility of cultivation in certain pockets of Eastern Ghats in Tamil Nadu. Spice India 5(7):19–20

Memory J (1956) 40% more flavour in improved vanilla process. Fd Engng (May 1975): 91, 93, 160

Nair KPP (2019) Intelligent Soil Management for Sustainable Agriculture – The Nutrient Buffer Power Concept. Springer Nature, Switzerland

Shadaksahri YG, Chandrappa HM, Umesha K (1996) Flowering and fruit set in vanilla (*V. planifolia*). J.Spices nd Aromatic Crops 5(2):148

Theodose R (1973) Traditional methods of vanilla propagation and their improvement. Trop Sci 15(1):47–57. In Proc. Spice Conference. Tropical Products Institute (TPI), London, pp 71–77